OXFORD BIOLOGY PRIMERS

Discover more in the series at
www.oxfordtextbooks.co.uk/obp

Published in partnership with the Royal Society of Biology

PROTEIN SCIENCE

OXFORD BIOLOGY PRIMERS

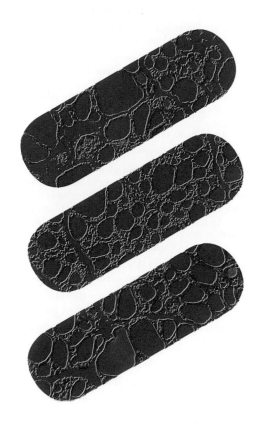

PROTEIN SCIENCE

Arthur Lesk

OXFORD
UNIVERSITY PRESS

Royal Society of
Biology

OXFORD
UNIVERSITY PRESS

Great Clarendon Street, Oxford, OX2 6DP,
United Kingdom

Oxford University Press is a department of the University of Oxford.
It furthers the University's objective of excellence in research, scholarship,
and education by publishing worldwide. Oxford is a registered trade mark of
Oxford University Press in the UK and in certain other countries

© Oxford University Press 2021

The moral rights of the author have been asserted

Impression: 1

All rights reserved. No part of this publication may be reproduced, stored in
a retrieval system, or transmitted, in any form or by any means, without the
prior permission in writing of Oxford University Press, or as expressly permitted
by law, by licence or under terms agreed with the appropriate reprographics
rights organization. Enquiries concerning reproduction outside the scope of the
above should be sent to the Rights Department, Oxford University Press, at the
address above

You must not circulate this work in any other form
and you must impose this same condition on any acquirer

Published in the United States of America by Oxford University Press
198 Madison Avenue, New York, NY 10016, United States of America

British Library Cataloguing in Publication Data

Data available

Library of Congress Control Number: 2021935064

ISBN 978–0–19–884645–1

Printed in Great Britain by
Bell & Bain Ltd., Glasgow

Links to third party websites are provided by Oxford in good faith and
for information only. Oxford disclaims any responsibility for the materials
contained in any third party website referenced in this work.

Dedicated to the memory of Cyrus Chothia

PREFACE

Whatever you may say about proteins, they are not boring. Proteins are responsible for a very large number and variety of biological activities: They are essential components of the structures of cells and tissues. They catalyse steps of metabolic pathways. They organize mechanisms of control over current cellular processes, and long-term developmental ones.

To carry out these functions, proteins present a great variety of three-dimensional structures. That these structures can be specified by the sequence of DNA relies on the fact that proteins fold spontaneously to the precise structures on which their biological activities depend.

In this book we shall explore the sources, mechanisms, and consequences of this infrastructure of biology.

This book has a companion web site: https://global.oup.com/he/lesk-obp1e. There you will find video clips of the structural pictures appearing in the text. The clips help in perceiving three-dimensional relationships. There are also sets of exercises, based on material from the chapters, which will be useful in testing your understanding of the subject matter. Compared with these, some of the Discussion Questions at the ends of the chapters are more far-reaching, and may require outside reading. References are suggested where appropriate.

It is assumed that the reader is familiar with the basics of general chemistry, organic chemistry, and biochemistry. I have provided explanations of specialized points as they arise.

The book is constructed for use in a classroom setting, under the guidance of an instructor. However, it would certainly be appropriate for anyone alone—in the classical situations: on a train or in a hotel room—or, today, perhaps quarantined from potential COVID-19 exposure—to read on his or her own.

CONTENTS

1 SETTING THE STAGE

Learning Objectives

- Understand the roles of proteins in the general context of life—proteins, together with many RNAs, are the 'executive branch'. You will recognize the place of proteins in Crick's **Central Dogma**: *DNA makes RNA makes protein*.
  - Some proteins have structural roles, both extracellular, such as hair; and intracellular, forming the cytoskeleton.
  - Some proteins, and protein-nucleic acid complexes, catalyse metabolic processes, including capturing energy from oxidation of glucose, and DNA replication and transcription.
  - Other proteins have regulatory roles: controlling metabolic processes, and gene expression.
- Appreciate the importance of the **spontaneous folding of proteins**—this is the point at which life makes the great leap from the one-dimensional world of DNA, RNA, and amino-acid sequences, to the three-dimensional world we inhabit.
- Understand how proteins explore the consequences of sequence variations, to evolve.
- Browse, and gain familiarity with, some of the many web sites that contain information about proteins.

We begin with a survey of the properties of proteins. Fundamental features of individual proteins are (1) the sequences of **amino acids** assembled into **polypeptide chains**, (2) the three-dimensional structures into which they fold, and (3) the functions they perform. Many pictures, using different representations, illustrate the essential characteristics and the great variety of protein structures. Many **databases** collect, and present on websites, our knowledge about proteins.

The diversity of protein structure and function has arisen through evolution. When sequence changes produce favourable variants—for a specific cell under specific conditions—selection can increase the frequency of a novel gene in a population. Conversely, sequence changes that produce protein dysfunction underly many diseases.

Biochemistry has isolated proteins and studied their individual properties. This is a necessary component of any approach to our ultimate goal: to understand the activities of proteins in living cells.

But if biochemists have taken cells apart, we now want to put them back together. It is well worth keeping the differences in these approaches clearly in mind. Control processes organize the functions of individual proteins into coherent cellular activity. This allows responses to changes in intracellular and extracellular environments, and implementation of inherent developmental programmes.

1.1   What are proteins, and what are they for?

Proteins are a family of macromolecules showing great variety in structure and function.

All proteins contain polypeptide chains: they are linear polymers of amino acids. There are 20 canonical amino acids, which appear in different orders in different proteins. The amino acids are linked by **peptide bonds** into a linear polymer. It is the sequences of the appearance of the amino acids along polypeptide chains that give proteins their great variety in structure and function.

The sequence of amino acids in each polypeptide chain is determined by the sequences of nucleotides in regions of the cell's **genome**. The human genome contains approximately 23,000 genes that code for proteins (exclusive of **antibodies**). Then, in turn, the **amino–acid sequence** of each protein determines its three-dimensional structure, and—thereby—its function. The understanding of protein function in terms of amino-acid sequence and three-dimensional structure is *the* fundamental challenge of protein science.

Study of the many known protein structures shows that underlying the great variety lurk recurring substructures, including but not limited to α–helices and β–sheets (see Chapter 2). These structural units form components of most proteins. Like 'lego bricks' they can be recombined in many different spatial configurations. Higher-level structural units, called **domains**, also recombine in different patterns, another mechanism for achieving structural diversity.

Different proteins adopt different, individual, three-dimensional structures, each structure appropriate for the assigned biological function. For example, enzymes contain **active sites**: clefts in their structures that specifically bind substrates, which interact with catalytic groups from the protein. Many other proteins function to regulate life processes, including governing the traffic through **metabolic pathways**, and control over **protein expression patterns**.

Some proteins contain multiple polypeptide chains. For example, human **haemoglobin** contains four polypeptide chains, two α chains and two β chains. These two types of chains are similar but not identical. The structure of this tetramer is required for haemoglobin to play its role in capturing oxygen efficiently in the lungs and delivering it effectively to the muscles.

In subsequent chapters, we shall expand on and develop these ideas. Chapter 2 treats protein structure in detail: the chemistry of the polypeptide chain, the nature of the 20 canonical amino acids, and the three-dimensional structures of proteins. Chapter 3 is oriented more towards experiment: methods for protein isolation and structure determination. Chapter 4 surveys protein functions. Chapter 5 builds on previous material to look at the relationships among natural proteins: the trajectories of evolutionary divergence that have allowed development of the many structures and functions that underly and implement the pageant of life.

1.2 The Central Dogma

Francis Crick's statement of the Central Dogma, in 1957, described the route of information transfer in cells:

DNA makes RNA makes protein

A cell's genome comprises the sequences of bases in DNA in the nucleus, and, separately, in organelles—mitochondria and chloroplasts. Regions of the genome are transcribed into **messenger** RNA. The messenger RNAs, after processing, are translated by the **ribosome** into the amino-acid sequences of proteins (see Figure 1.1). In eukaryotes, many protein-coding genes are discontinuous. They contain **exons**—regions translated into amino-acid sequences—and **introns**—non-coding regions. Processing of messenger RNA splices out introns. By including different combinations of exons in

Fig. 1.1 Implementation of the Central Dogma, in structural terms. Regions in the DNA sequence (red and blue, top) encode amino-acid sequences of proteins. In transcription, RNA polymerase creates a single-stranded messenger RNA (mRNA) molecule with a copy of a region of one of the strands of the DNA (mRNA shown in purple). (In eukaryotes, in many cases, processing is necessary to ensure that the mRNA corresponds to the component of the DNA sequence that encodes the protein.) In translation, mRNA passes through the ribosome—which acts as a 'tape reader'—sequentially to synthesize the polypeptide chain (pink discs), directed by the sequence of bases in the mRNA.

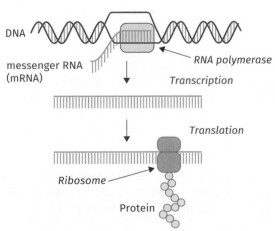

the result, or variable splicing, a single region of DNA can be translated to several different proteins.

Crick wished to emphasize that DNA sequences inform RNA sequences, and RNA sequences inform amino-acid sequences, but that the information does not travel back directly from protein to RNA to DNA. We now know that some viruses, notably HIV-1 that causes AIDS, can reverse transcribe the viral RNA genome into the host's nuclear DNA. (This is one of the reasons that HIV-1 infections are difficult to treat—even if drugs clear the virus from the blood, its genome lies in wait in the patient's DNA, from which it can reemerge.) Reverse transcription is an exception to, or an extension of, the Central Dogma.

Also, although information cannot flow *directly* from protein to RNA to DNA, changes in proteins arising from mutations in DNA can affect the DNA through evolutionary selection. Selection can enhance the frequency within a population of a mutation that confers advantageous features on some protein. Conversely, if a mutation destroys the activity of an essential protein, the cell will not survive, and the mutant DNA will be eliminated from the population.

1.3   The spontaneous folding of proteins

The Central Dogma takes us as far as amino-acid sequences. But to carry out their functions, most proteins must adopt an exquisitely precise three-dimensional structure. (See The bigger picture 1.1.)

Protein structures in three-dimensions

Understanding the biological roles and activities of proteins at the molecular level requires knowing their structures. We now have detailed structures—the actual $x, y,$ and z coordinates of every atom—for over 100,000 proteins. We use computer graphics to create pictures of proteins. Several different types of representation are useful. Figure 1.3 shows four pictures of the structure of a small protein, acylphosphatase. Being able to manipulate such pictures

The bigger picture 1.1
From one dimension to three

The spontaneous folding of proteins to form their native states is the point at which nature makes the giant leap from the one-dimensional world of gene and protein sequences to the three-dimensional world that we inhabit. There is a paradox—it may be regarded as ironic that the translation of RNA into protein is *logically* very simple—the genetic code specifies the translation of triplets of nucleotides (called **codons**) into amino acids. But the implementation of this directed polymer synthesis requires immensely complicated machinery, including the ribosome. In contrast, the way the amino-acid sequence specifies the three-dimensional structure is logically extremely complicated—but it happens spontaneously (see Figure 1.2).

Fig. 1.2 'A most ingenious paradox': the translation of DNA sequences to amino-acid sequences requires immensely complicated machinery. The folding of the polypeptide chain into a three-dimensional structure takes place spontaneously.

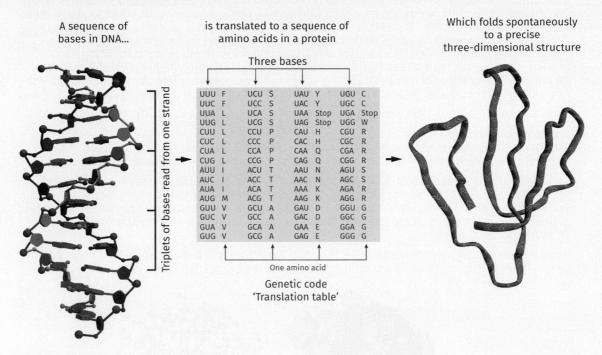

A sequence of bases in DNA...

is translated to a sequence of amino acids in a protein

Which folds spontaneously to a precise three-dimensional structure

Three bases

Triplets of bases read from one strand

UUU F	UCU S	UAU Y	UGU C
UUC F	UCC S	UAC Y	UGC C
UUA L	UCA S	UAA Stop	UGA Stop
UUG L	UCG S	UAG Stop	UGG W
CUU L	CCU P	CAU H	CGU R
CUC L	CCC P	CAC H	CGC R
CUA L	CCA P	CAA Q	CGA R
CUG L	CCG P	CAG Q	CGG R
AUU I	ACU T	AAU N	AGU S
AUC I	ACC T	AAC N	AGC S
AUA I	ACA T	AAA K	AGA R
AUG M	ACG T	AAG K	AGG R
GUU V	GCU A	GAU D	GGU G
GUC V	GCC A	GAC D	GGC G
GUA V	GCA A	GAA E	GGA G
GUG V	GCG A	GAG E	GGG G

One amino acid

Genetic code
'Translation table'

Discussion Questions

1. There is evidence that proteins begin to fold as they emerge from the ribosome. This suggests the general picture that the first part of the protein to be synthesized acts as a nucleus for folding, and subsequently-synthesized regions then fold around this nucleus. But there are protein structures known for which the *last* segment synthesized sits at the centre of the interior of the protein structure, with the rest of the chain folded around it. Can you reconcile these observations with the general picture?

2. Amino-acid sequences have evolved to fold spontaneously into protein structures. The requirements for this are: (a) the folded state achieved has a greater stability than other possible conformations of the chain, and (b) there is a low-energy 'pathway' from the un-folded state to the folded state, along which there are no barriers that would prevent attainment of the correct folded state.

   Suppose you had a protein structure in which the two ends of the chain were nearby in space. This means that you could synthesize a cyclically-permuted polypeptide chain, by forming a bond between the two original ends, and breaking the chain at some other point. Would you expect the cyclically-permuted chain to have at least close to the same minimum-energy conformation as the original chain? How might you expect the folding behaviour of the cyclically-permuted chain to compare with that of the original chain?

Fig. 1.3 Representations of the structure of acylphosphatase. (Acylphosphatase catalyses the reaction: $R(C=O)OPO_3^{2-} + H_2O \rightarrow RCOO^- + PO_4^{3-} + 2H^+$, where R may be any of several acyl groups.)

(a) The most detailed representation of the structure, including mainchain and sidechains. Each atom (except hydrogens) is shown as a sphere of its Van der Waals radius. (Grey = carbon, red = oxygen, blue = nitrogen, yellow = sulphur.) The Van der Waals radius of an atom is the minimal allowable distance from an atom to any other atom to which it is not chemically bonded, at the point of collision. This figure demonstrates why simplification is necessary to produce an intelligible picture of even a small protein. However, this representation does give a clear sense, which simplified representations often do not, that the structure of a globular protein is a compact and dense assembly of atoms. You could not shoot a water molecule through the protein, as you might think from the other panels of this figure.

(b) is also an 'all-atom' picture, showing the structure in 'ball-and-stick' representation.

(c) contains a simplified representation, tracing the course through space of the mainchain. Sidechains are not shown. Chevrons indicate the local chain direction. Can you visually trace the entire chain? Start with the free end at the lower left. (The video clips in the online resources will be helpful.)

Two regions at the front of the picture have the form of *helices*—like classic barber poles—with their axes almost vertical in the orientation shown. (They are coloured scarlet.) The structure also contains five regions in which the chain is extended, almost a straight line; these too are approximately vertical in orientation in the figure. These regions interact laterally to stabilize their assembly into a *sheet*. (They are coloured green.)

In (d), helices and strands are represented by 'icons': helices as cylinders and strands of sheet as large arrows. The cylinders are translucent, so that you can see the arrows behind them.

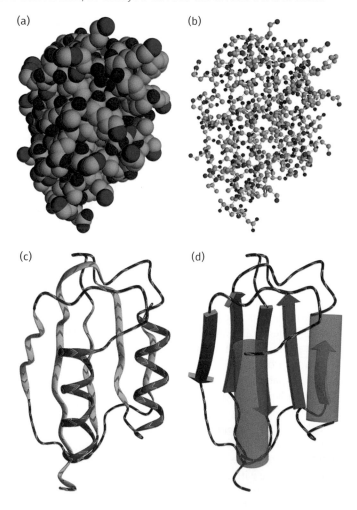

interactively on a screen greatly enhances perception of three-dimensional relationships (see online resources associated with this book).

1.4 Proteins and genomics

Each element of the Central Dogma has given rise to an 'omics discipline. The determination and analysis of genome sequences is **genomics**. For the most part, the genome sequence must be determined only once per individual, as most cells in an organism have the same DNA. (One exception is cancer genomics, in which the *differences* in DNA sequences between normal and tumour cells in the same individual are studied intensively.) The differences in genome sequences between different individuals of the same species are generally small, although they may have large phenotypic consequences.

Transcriptomics is the study of transcribed RNA sequences, not limited to messenger RNAs.

Proteomics is the *direct* study of proteins in a cell or organism. Whereas genome sequencing gives us general information about the potential proteins encoded, to discover precisely how this potential is implemented requires study of proteins themselves.

 Key point

The steps of the Central Dogma—DNA → RNA → protein—correspond to three thriving areas of research: genomics, transcriptomics, and proteomics.

1.5 Proteomics

What do we need to know about proteins that genome sequences do not tell us? Proteomics involves the measurement of:

- The inventory of proteins—which proteins, and how much—and how this inventory varies within cells, between tissues, and in different states—for instance yeast growing under aerobic or anaerobic conditions.

- Details about proteins not available from the genome sequence, including

    - first and foremost: the three-dimensional structures of expressed proteins

    - **splice variants**, choices for translation of different combinations of segments of a gene

    - assembly into **multimeric structures**

    - post-translational modifications

    - **cofactors** that are an integral and non-transient part of protein structures

    - protein functions

    - interactions between different proteins

    - subcellular localization of proteins

    - organization of protein functions into networks, of metabolic pathways, or **regulatory networks**

Methods for investigating these questions include high-throughput technologies such as mass spectrometry, and gel electrophoresis.

Protein mass spectrometry

Mass spectrometry involves the measurement of the trajectories of charged particles in a magnetic field. The magnetic field bends the paths of the particles into circles perpendicular to the magnetic field. The radius of the circle, which is what is measured, determines the charge/mass ratio of the molecular ion.

Technical problems include: (a) getting the proteins into the gas phase without destroying them, and (b) converting neutral particles to ions. Two solutions are electrospray ionization (ESI), and matrix-assisted laser desorption/ionization (MALDI). In electrospray ionization, a high voltage is applied to a liquid, producing an aerosol containing the protein solutes. Matrix-assisted laser desorption/ionization involves embedding proteins in a solid matrix and applying pulses of laser light.

A common approach to mass spectral analysis of a protein sample is to digest the protein with trypsin, an enzyme that cleaves polypeptide chains at specific positions. The fragments produced typically have average molecular weights of 700–1500 (about 6–14 amino acids). The distribution of sizes of trypsin-produced peptides is individual to each protein. Databases of sizes of predicted tryptic fragments are easily derivable from amino-acid sequences. This allows identification of a protein from the mass spectrum of its tryptic fragments. However, the size distribution of tryptic fragments of a *novel* protein will not appear in the database.

Two-dimensional gel electrophoresis

Given a heterogeneous mixture of proteins harvested from a source, such as a lysate of a bacterial culture or a mammalian liver, the components can be fractionated by two-dimensional gel electrophoresis (see Figure 1.4).

Fig. 1.4 Two-dimensional gel electrophoresis is a two-step process. (a) First the proteins move, in *one* dimension, in an electric field. A pH gradient is established, such that different points in the path have different values of pH. Because proteins contain titratable groups, they will have different charges at different pH values of the medium. When charged they will migrate under the influence of the electric field. Some, initially bearing a positive charge, will migrate towards lower pH; others, initially bearing a negative charge, will migrate towards higher pH. But, for each protein, at some pH the net charge will be zero. That is called the **isoelectric point (pI)** of the protein. When each protein reaches the point in the gradient where the pH is equal to its isoelectric point, it will no longer move. (The electric field does not exert a force on uncharged molecules.) This is called **isoelectric focussing (IEF)**. The effect is to sort out proteins according to their isoelectric points. (This is related to the charge on the protein at pH7 but is not quite the same property.) (b) However, there may be many proteins in the original mixture with the same or similar isoelectric points. The isoelectric focussing step will not separate them. Exposure of the sample to **sodium dodecyl sulfate (SDS)**, a detergent, unfolds the proteins, and coats them with a uniform negatively-charged layer. The results of the first, isoelectric focussing step, are applied to an edge of a *two-dimensional* flat polyacrylamide gel. Proteins so treated migrate in electrophoresis according to their molecular weight. This separates, into different components, the proteins contained in each band arising in the first, IEF, step. The positions of the 'spots', and estimates of their concentrations, are available by staining the gel. ((a) and (b) are schematic.) (c) To identify the protein or proteins contained in any spot in the two-dimensional gel, it is possible to form tryptic peptides, and apply mass spectrometry.

Fig. 1.4 (*Continued*)

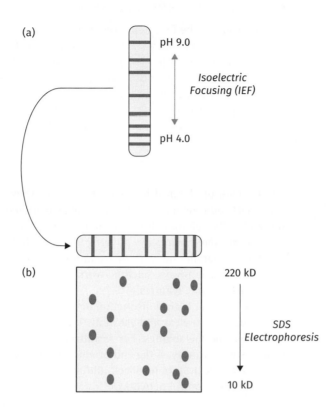

(a)

pH 9.0

Isoelectric Focusing (IEF)

pH 4.0

(b)

220 kD

SDS Electrophoresis

10 kD

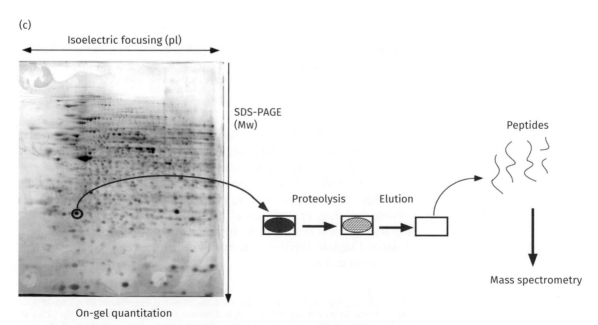

(c)

Isoelectric focusing (pI)

SDS-PAGE (Mw)

On-gel quantitation

Proteolysis   Elution

Peptides

Mass spectrometry

(Part (c) Reprinted from Journal of Proteomics, 73/11, Thierry Rabilloud, Mireille Chevallet, Sylvie Luche, Cécile Lelong, Two-dimensional gel electrophoresis in proteomics: Past, present and future, Pages 2064–77, Copyright (2010), with permission from Elsevier.)

💡 **Key point**

To summarize: The first step, isoelectric focussing, separates a mixture of proteins into components with the same isoelectric points. The second step, polyacrylamide gel electrophoresis (PAGE), further separates the mixture according to molecular weight. Each spot on the resulting gel contains a unique protein, or possibly several proteins, that have specific values of *both* isoelectric point and molecular weight.[1]

Fig. 1.5 A generic metabolic pathway, with the end product E acting as an inhibitor of the enzyme that catalyses the first step, the conversion of A to B. The 'T' symbol conventionally signifies repression. The effect is to shut the pathway down if the end product E is present in adequate amounts.

1.6   Regulation of protein activity

The triumph of classical biochemistry was to show that proteins isolated from cells could retain specific and independent activity after purification. But within the cell these functions must be integrated and regulated. A network of *metabolic pathways* governs the possible interconversions of metabolites. A parallel network of *regulatory interactions* controls the traffic through these pathways. Each network—metabolic or regulatory—has both static and dynamic features.

There is a clear logical distinction between metabolic and regulatory networks, but not so clear a physical distinction, as signalling molecules may themselves be metabolites. For instance, **feedback inhibition** may act to shut down a pathway, if the end product is present in adequate concentration (see Figure 1.5). In feedback inhibition a molecule participates in *both* metabolic and control networks.

💡 **Key point**

Think of **intermediary metabolism**, catalysed by enzymes, as the 'smokestack industries' of the cell, and the regulatory systems, including signal transduction and expression control, as the 'silicon valley'.

Feedback inhibition affects the activity of proteins already present and active. In addition, living things regulate the synthesis of proteins encoded in their genomes. To achieve this, transcription and translation must also be dynamic—to produce the right amount of the right protein at the right time at the right place. In this way, cells can respond to stimuli by altering their physiological state, or even their physical form. The driving force, for these changes of protein expression profile, may be changes in the environment, or internal signals directing different stages of the cell cycle, or developmental programmes.

---

[1]Occasionally, during a lecture, I have carried out the following 'experiment'. First, I ask all male students to step to the left side of the room, and females to step to the right side. Then I ask all students wearing clothing showing the logo of the university to step to the front of the room, and the others to move to the back of the room. Now there are four groups. By a succession of such steps it would be possible to separate out a single individual.

Well-studied examples include:

1. The replacement of glucose by lactose in the medium can trigger transcription of the lactose **operon** in *E. coli*. (An operon is a series of successive genes under control of a single promoter, and co-transcribed.)

2. Transcription of another operon, encoding enzymes for the biosynthesis of the amino acid tryptophan, will be repressed if tryptophan is present in adequate concentrations.

3. A consequence of regulation at a higher level: Human cells may differentiate characteristically in different tissues; for instance to turn into neurons, sprouting dendrites and an axon, and expressing tissue-specific or even cell-specific proteins.

Cells overlook no opportunity to exert control. The Central Dogma of DNA → RNA → protein suggests several possible leverage points for regulation of protein activity. Figure 1.6 indicates some of the control mechanisms that apply to different steps. It is approximately true that mechanisms that apply at the levels of expressed proteins have faster effects than those that control gene expression. When Jacques Monod investigated **diauxy** in yeast in the 1940s he noted in several cases a 'lag phase' as the cells converted between alternative active metabolic pathways. (Diauxy means 'double-growth'.) The cells needed time to 'retool' by changing patterns of gene expression.

In prokaryotes, a specific focus of transcriptional regulation is at or near the binding site of RNA polymerases to DNA, before (5′ to) the beginning of the gene. **Repressors** can turn off transcription by occluding the binding site, blocking RNA polymerase activity (see Figure 1.1). In contrast,

Fig. 1.6 Regulation of protein activity involves many mechanisms, active at different stages of protein expression and activity.

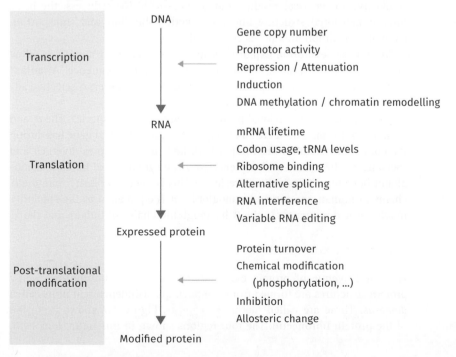

promoters can actively recruit polymerases through cooperative binding, along with polymerase, to a site on the DNA.

In eukaryotes, gene regulation is more complex. Transcription regulators may bind to DNA at positions proximal to the gene as in many cases in prokaryotes, but also at remote sites. Regulatory interactions also govern the expression of other transcription factors, producing networks of regulatory activity. Eukaryotic control networks show far greater complexity, in both their logic and their dynamics, than those of viruses or prokaryotes.

The gift of complexity is robustness. Eukaryotic control networks show an ability to reprogramme themselves, to respond to stimuli by changing cell state. The source of robustness appears to be redundancy. Yeast *(Saccharomyces cerevisiae)* has about 6,000 genes. Under 'normal'—non-stress—conditions, about 80% of them are being expressed. It is also true that yeast can survive approximately 80% of single-gene knockouts. Many expressed genes must be redundant, and the redundancy is the source of the robustness.

Fig. 1.7 A fragment of fibronectin, a modular protein, showing four tandem domains.

1.7   Protein evolution

Comparison of amino-acid sequences and three-dimensional structures of corresponding proteins from different species—for instance, human and horse haemoglobin—show evolutionary divergence. The degree of divergence in corresponding proteins between species usually mirrors the degree of divergence of the species in the standard evolutionary tree.

For instance, the α chains of human and horse haemoglobin (two mammals) have 123 identical amino acids, out of a total of 141—87%. The α chains of human and turkey haemoglobin (mammal v. bird) have lower similarity: 99 identical residues out of 141—70%. Nevertheless, the basic three-dimensional structure and the function—binding and transporting oxygen and carbon dioxide—is conserved.

We thus see evolution exploring sequence space. Point mutations allow evolution to grope around in the vicinity of a parent sequence. A mutant protein may take over a population either because it confers a selective advantage; or, in some cases, at random, by genetic drift.

Often we see several related proteins *within* the same species. The α and β chains of haemoglobin are an example. A single ancestral gene has during the course of evolution been duplicated, and the two copies diverged and specialized. The amino-acid sequences of α and β chains of human haemoglobin have only 43% identical residues. This is lower similarity than the α chains of human and turkey haemoglobin; it is equivalent to the similarity in amino-acid sequences of the α haemoglobin chains of human and shark.

Domain recombination

A second avenue of protein evolution is **domain recombination**. Many protein structures are formed from compact, quasi-independent units called *domains*. These are called modular proteins. Figure 1.7 shows a portion of the protein fibronectin. The four regions shown in this figure look as if, were they clipped out of the whole protein, they would fold independently to the same structure. This defines them as domains.

Distinguish modular proteins, which contain multiple independent substructural units within a single polypeptide chain (fibronectin), from oligomeric proteins, composed of two or more separate polypeptide chains forming a coherent complex (haemoglobin).

Fibronectin shows a repeat of the same structural type of domain. Many proteins comprise sets of structurally and functionally diverse domains. By recombining different domains in different orders, proteins have easy access to different combinations of structures and functions. An alternative approach, to design a novel structure and function 'from scratch', is harder. The system would have to work out an amino-acid sequence that would generate the required fold. Why not just reuse and adapt 'off-the-shelf' components?

 Key point

Pathways of protein evolution include: (1) *local changes*, such as point mutations, and (2) *larger scale recombinations*: 'mixing and matching' of domains.

1.8   Protein dysfunction, and disease

A variety of types of mutation can change the amino-acid sequences of expressed proteins, or alter the regulation of their expression. In many cases perturbation, or complete loss, of function can have clinical consequences. Protein dysfunction is the cause of many diseases.

Linus Pauling first recognized a disease with its origin at the molecular level. Sickle-cell disease is the result of a single-site mutation in haemoglobin. A normal glutamic acid (Glu) is replaced by a valine (Val), on the surface of the protein. The sidechain of glutamic acid is charged; that of valine is non-polar. The non-polar sidechain of valine has a thermodynamically unfavourable interaction with water, and favourable interaction with other non-polar sidechains. The effect of the mutation is to create a 'sticky patch' on the protein surface. As a result, the mutant haemoglobin is prone to aggregate, in the deoxy state. (The small structural changes between oxy and deoxy states are sufficient to prevent the mutant haemoglobin from aggregating in the oxy state.) The symptoms include both long-term effects, and crises involving localized pain; sometimes precipitated by oxygen demands during exercise, which converts more haemoglobin to the deoxy state.

It is estimated that approximately 14,000 people in the United Kingdom suffer from sickle-cell disease. The mutant allele is much more prevalent in individuals of African or Afro-Caribbean origin.

Why was the mutant allele not removed by selection? The sickle-cell trait provides protection against malaria. Heterozygotes, having one allele for normal haemoglobin and one for the mutant, are clinically near-normal with respect to haemoglobin aggregation, and therefore pay a relatively small price compared to the advantages of malaria protection conferred.

There are many genetic diseases. Some, like sickle-cell disease, are the result of single-site mutations. However, many other changes to DNA can have clinical consequences. These include insertions and deletions, changes in copy number (for instance, gene duplications), inversions, and translocations.

Understanding the association of diseases with changes in protein functions supports clinical applications:

- *Protein replacement* In some cases, absence of a functional protein can be overcome by providing a working protein. Insulin for diabetes, and blood-clotting factors for haemophilia are two well-known examples.

- *Genetic counselling* In many cases, the disease condition is genetically recessive. In these cases heterozygotes may be clinically normal themselves, but be *carriers* of the mutant gene. If a man and woman are contemplating marriage and children, genetic sequencing can inform them whether they are carriers of an abnormality, and the likelihood of their children's showing a disease characteristic of homozygotes. This used to be applied primarily in cases where family histories raised warning flags, but with large-scale sequencing, carrier detection will become automatic.

- *Neonatal screening* Some conditions are detectable at birth (and even before!) allowing for immediate clinical intervention. For example, **phenylketonuria** is a disease arising from dysfunction of the enzyme phenylalanine hydroxylase, that converts the amino acid phenylalanine to tyrosine. Untreated, the accumulation of phenylalanine and its by-products leads to developmental defects, including intellectual disability. Approximately 1 in 10,000 babies born in the UK has phenylketonuria. Screening for the condition allows treatment to enable normal development. In the past, the treatment involved a lifelong low-phenylalanine diet. Recently a treatment based on a bacterial enzyme, phenylalanine ammonia-lyase, has been approved in the EU and the US.

- *Genetic engineering* The relatively new technique, CRISPR, allows editing of a genome to revert a mutation; for instance, changing the haemoglobin mutant in sickle-cell disease back to the normal sequence. The clinical potential of CRISPR for genetic diseases is immense.

 Key point

Many diseases arise from dysfunction of particular proteins. Knowing the genes for the proteins implicated permits *warning* (genetic counselling), *detection* (neonatal or even foetal screening), and *genetic engineering* to repair the mutation (using CRISPR).

1.9   Databases and web sites containing information about proteins

Through a wide variety of experimental techniques, we are accumulating very large amounts of information about proteins. Databases are essential to assemble this information in useful form.

A database must:

- collect data
- curate the data: that is, perform quality checks, and record provenance (the source of the data: who measured it, and with what technique?)
- impose structure on the data so that it is readily accessible by search engines and other information-retrieval tools ('A database without effective means of access is merely a data graveyard.')
- provide access to the database—now this is always done through a web site
- provide links to other databases containing related information.

 Key point

Although creation and maintenance of a database is a speciality, all biologists and clinicians need to know how to use them.

Database organization in molecular biology requires several types of abilities, spanning biology and computing. A new professional speciality has arisen to master and apply these abilities.

Formerly, different types of information resided in different databases. Each database was organized by people expert in the relevant experimental technique, capable of specialized informed judgement. There has been a trend to combine databases into comprehensive collections of different types of information. This was stimulated partly by the need to ask questions requiring access to different types of data. For instance, consider the question: 'For which proteins of known structure involved in diseases of purine biosynthesis in humans, are there related proteins in yeast?' We require *simultaneous* access to information about amino-acid sequence, three-dimensional structure, metabolic pathways, and disease. Such questions are difficult to address if the different sectors of information occupy different databases.

A few (of the many) major databases containing information about proteins are (see Table 1.1):

- *The UniProt Knowledgebase (UniProtKB)*: This database contains entries for individual proteins, collecting information from the scientific literature, and results of computational analysis. Each entry includes but is by no means limited to information about the amino-acid sequence, three-dimensional structure, and protein function; each entry is extremely rich in links to other databases with information about the protein.

Table 1.1 Home pages of databases containing information about proteins

Database	Home Page
The UniProt Knowledgebase (UniProtKB)	www.uniprot.org
The World-Wide PDB (wwPDB)	www.ebi.ac.uk/pdbe/
	www.rcsb.org
	pdbj.org
SCOP	scop.mrc-lmb.cam.ac.uk
CATH	www.cathdb.info
DIP	dip.doe-mbi.ucla.edu
IntAct	www.ebi.ac.uk/intact/
MINT	mint.bio.uniroma2.it
STRING	string-db.org
MetaCyc/BioCyc	metacyc.org
	biocyc.org
BRENDA	www.brenda-enzymes.org
Reactome	reactome.org
UniPathway	www.uniprot.org/database/DB-0170
KEGG	www.genome.jp/kegg/
OMIM	omim.org
OMIA	omia.org

- *The Protein Data Bank (PDB)*: This database collects information about the three-dimensional structures of biological macromolecules. Each entry contains the coordinates of a solved structure of a protein, nucleic acid, or protein-nucleic acid complex, plus small molecules bound, information about *how* the structure was determined, and related information.

- *SCOP* and *CATH*: Databases showing structural relationships among known structures.

- Databases *DIP*, *IntAct*, *MINT,* and *STRING* cover protein-protein interactions.

- *MetaCyc/BioCyc, BRENDA (BRaunschweig ENzyme DAtabase), Reactome, UniPathway,* and *KEGG (Kyoto Encyclopaedia of Genes and Genomes)* treat enzymes and metabolic pathways.

- *Online Mendelian Inheritance in Man (OMIM)* is a catalogue of human genetic disorders, describing the clinical consequences of mutations.

- *Online Mendelian Inheritance in Animals (OMIA)* is a corresponding catalogue of genetic disorders in 250 (non-human) animal species.

A 2017 article by Chen, Huang, and Wu (2017) (see Further Reading) gives thorough descriptions of these and other databases, and provides 'clickable' links (this is why it would be convenient to access it on a computer rather than on paper). The reader is urged to explore some of these databases, initially only to get a sense of what they contain and how to use them.

Summary Points

- Proteins are created through the steps of Crick's Central Dogma: Regions of the genome, transcribed to RNA, are translated by ribosomes to the amino-acid sequences of proteins.
- Proteins fold *spontaneously* to native structures: this is the point where life makes the leap from the one-dimensional world of DNA, RNA, and amino-acid sequences, to three-dimensions.
- Proteins provide a variety of functions, including structural components of cells and tissues, and the catalytic activities of enzymes. Regulation of protein expression and activity—to a large extent, *by* proteins—is essential in organizing the activities of cells.
- Proteins evolve: the two main mechanisms of protein evolution are mutations at individual sites that allow local exploration of sequence space, and 'mixing and matching' of different domains. Conversely, mutations that lead to protein dysfunction are the cause of many diseases.
- Many web sites give access to databases that collect, curate, and organize what we know about proteins.

Further Reading

Alberts, B. (1998). The cell as a collection of protein machines: preparing the next generation of molecular biologists. Cell 92, 2914.
A thoughtful assessment of the field, albeit from some time ago, by an individual at the centre of the developments.

Chen, C., Huang, H. and Wu, C.H. (2017). Protein bioinformatics databases and resources. Methods Mol. Biol. 1558, 3–39.
A comprehensive survey of major protein databases, and discussion of challenges and opportunities for future developments.

Cobb, M. (2017). 60 years ago, Francis Crick changed the logic of biology. PLoS Biol 15:e2003243.
Provides historical context and a tribute to Crick's always profound insight.

Doudna, J.A. and Sternberg, S.H. (2017). A Crack in Creation: Gene Editing and the Unthinkable Power to Control Evolution (Boston, MA: Houghton Mifflin, Harcourt).
Discussion of recent breakthroughs in gene editing, and the scientific, clinical, and social consequences.

Eaton, W.A. (2020). Hemoglobin S polymerization and sickle cell disease: A retrospective on the occasion of the 70th anniversary of Pauling's Science paper. Amer. J. Hematology 95, 205–11.
A masterful case study of a disease important both to a large number of patients, and to the history of molecular biology.

Eng, J.K., Searle, B.C., Clause, K.R. and Tabb, D.L. (2011). A face in the crowd: Recognizing peptides through database search. Mol. Cell. Proteomics 10, R111.009522.
Discussion of databases supporting peptide identification via mass spectrometry.

Lesk, A.M., Bernstein, H.J. and Bernstein, F.C. (2008). Molecular graphics in structural biology. In: Computational Structural Biology, Methods and Applications, M. Peitsch, and T. Schwede, eds. (Singapore: World Scientific Publishing) 729–70.
A review of molecular graphics and its applications.

Misra, B.B., Langefeld, C., Olivier, M. and Cox, L.A. (2019). Integrated omics: tools, advances and future approaches. J. Mol. Endocrin. 62, R21–R45.
A survey of the 'omics disciplines explaining how they fit together.

Discussion Questions

1.1 Many people believe that there was an early era of life based on RNA, before proteins existed. What functions of proteins would be relatively easy to accomplish with RNA molecules? What functions of proteins would be relatively difficult to accomplish with RNA molecules? What is the current status of the RNA-world hypothesis? (Orgel, L.E. (2004). Prebiotic chemistry and the origin of the RNA world. Crit. Rev. Biochem. Mol. Biol. 39, 99–123; Le Vay, K. and Mutschler, H. (2019). The difficult case of an RNA-only origin of life. Emerging Topics in Life Sciences 3, 469–75; Ball, P. (2020). Flaws in the RNA world. Chemistry World https://www.chemistryworld.com/opinion/flaws-in-the-rna-world/4011172.article).

1.2 What successes has the genetic-engineering technique CRISPR had in curing metabolic diseases? What types of constraints on its use are under discussion? What are the concerns raised that might argue for limitations of its use? (Doudna and Sternberg, see Further Reading; https://ghr.nlm.nih.gov/primer/genomicresearch/genomeediting and articles cited there; Ledford, H. (2020). Quest to use CRISPR against disease gains ground. Nature 577, 156.)

2 PROTEIN STRUCTURE

Learning Objectives

- Understand the principles of construction and design of proteins, at every level from amino acids up to complete structures. The variety of physicochemical properties among the 20 standard amino acids, and the ability of the polypeptide chain to fold up into different three-dimensional structures, contribute to proteins' great versatility.

- Recognize how amino acids form peptide bonds, to create a polymer chain. The chemical structure of the backbone, or main-chain, is the same in different proteins. The sequences of **sidechains** determine the three-dimensional structures and, thereby, the functions, of different individual proteins.

- Distinguish the properties of the native and denatured states. Each protein can form a native state, folding into a compact three-dimensional structure dictated by its amino-acid sequence. In contrast, the denatured state arises when conditions of temperature or solvent break up the native state. In denatured states the interresidue interactions are lost, the structure is no longer compact, and function is lost. In dilute solution, denatured proteins can recover their native states if brought back to normal conditions.

- Be able to recognize—in a picture of a protein structure—common substructures, including:
  - secondary structures, α–helices and β–sheets
  - supersecondary structures, combinations of α–helices and strands of β–sheets characterizing local regions of the chain
  - domains, larger substructures that show similar folding patterns in different structural contexts.

- Recognize that corresponding proteins in different species diverge in amino-acid sequence. However, families of related proteins, for instance haemoglobins, often retain structures and functions.

• Know that proteins can interact to form complexes. Many are natural and contribute to normal, healthy, function. Others, especially aggregates of denatured or misfolded proteins, can cause disease.

Living things achieve great variety in their proteins by assembling different sequences of amino acids, into long linear polymer chains called polypeptides. The amino-acid sequence of each protein determines its three-dimensional structure, and function. This chapter introduces the basic chemistry of proteins, and illustrates some of the salient features of their three-dimensional structures.

2.1   Proteins are formed of polypeptide chains

Proteins are linear polymers of amino acids (see Figure 2.1(a)). All proteins have in common the backbone, or main-chain, structure of the polypeptide chain (see Figure 2.1(b)).

Notice that the first amino acid retains a free protonated amino group, and the last amino acid retains a free carboxylate group. These are called the N-terminus and the C-terminus of the polypeptide chain. We conventionally write amino-acid sequences in the N-terminus to C-terminus direction. Within a structure we can speak of the local direction of a region; for instance in Figure 1.3(d) the large arrows point in the N to C direction.

Fig. 2.1 (a) The chemical structure of the amino acids. Each amino acid has a central carbon, to which is attached a hydrogen, a carboxylate group, an amino group, and a sidechain. The central carbon is called the Cα. The amino acids that make up proteins are chosen, in almost all cases, from a standard set of 20. For most amino acids the central carbon is asymmetric, and the amino acid could adopt two mirror-image conformations. The figure is drawn to suggest that the COO^- and NH_3^+ groups are in front of the central carbon, and that the H and the sidechain are behind. Almost all natural amino acids adopt this L conformation at the Cα. (b) To form the polypeptide chain, the ribosome catalyses the reaction of the COO^- of one amino acid with the NH_3^+ of another,

eliminating water to form the *peptide bond*. $\begin{array}{c} H \\ | \\ C-N \\ || \\ O \end{array}$ In this case the sidechain of the first amino

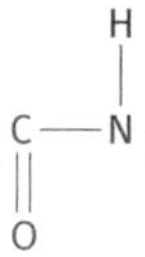

acid is CH_3, and that of the second is CH_2OH. (After you have studied Figure 2.3: which amino acids are they?)

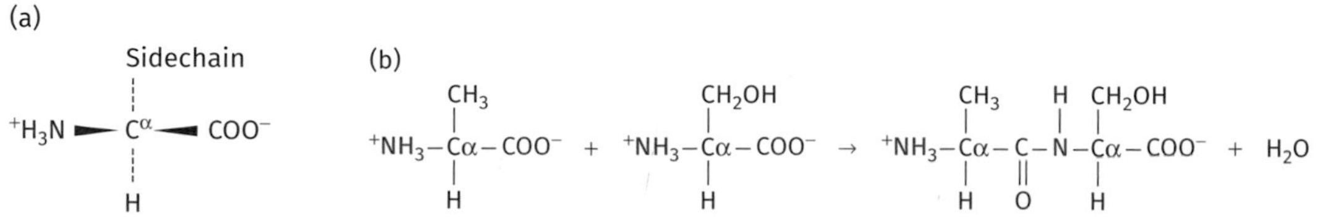

(a)

Fig. 2.2 Proteins contain a series of residues linked via peptide bonds into a polypeptide chain with a repetitive mainchain, and a variable sequence of sidechains S_i attached to successive residues.

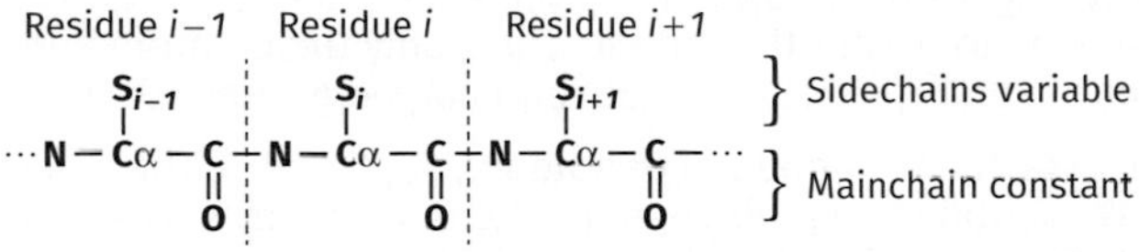

Residue $i-1$ Residue i Residue $i+1$

$$\cdots N - C\alpha - \underset{\underset{O}{\|}}{C} + N - C\alpha - \underset{\underset{O}{\|}}{C} + N - C\alpha - \underset{\underset{O}{\|}}{C} - \cdots$$

with S_{i-1}, S_i, S_{i+1} attached.

} Sidechains variable

} Mainchain constant

Attached to the main chain at regular intervals are sidechains (see Figure 2.2). Each sidechain is, almost always, chosen from a set of 20 possibilities. The sequences of sidechains gives different proteins their individuality, and account for the variety of their structures and functions.

One may think of a protein chain as similar to a string of Christmas-tree lights: the wire, with sockets for bulbs attached, is the backbone. This backbone will be the same in each household; although, like proteins, the chains may have different lengths. The sequence of colours of the bulbs is variable. (But, for strings of lights the sequences of colours do NOT automatically determine the spatial conformation of the wire!)

Key point

The mainchain has the *same* structure in all proteins. This means that the mechanism of peptide bond synthesis by the ribosome is amino-acid sequence independent. The sequence of sidechains differs among different proteins.

2.2   The sidechains

Each residue in a protein has a sidechain chosen from a set of 20 possibilities. This is the 'cast of characters':

Amino acids with non-polar sidechains							
G	glycine	A	alanine	P	proline	V	valine
I	isoleucine	L	leucine	F	phenylalanine	M	methionine
Amino acids with polar sidechains							
S	serine	C	cysteine	T	threonine	N	asparagine
Q	glutamine	Y	tyrosine	W	tryptophan		
Amino acids with positively-charged sidechains							
K	lysine	R	arginine	H	histidine		
Amino acids with negatively-charged sidechains							
D	aspartate	E	glutamate				

It is common, especially in writing sequences, to designate each amino acid by a one-letter abbreviation of its name.

Figure 2.3 shows the chemical structures of the twenty amino acids. The side-chains attached to each residue of a protein show a variety of sizes and physico-chemical properties, affording proteins great versatility of structure and function:

Fig. 2.3 (a) Structures of the sidechains (hydrogens not shown). For glycine the main-chain atoms are shown; the sidechain of glycine is a hydrogen atom. For other amino acids, only the $C\alpha$ atom and side-chain are shown. Positively charged groups (e.g. NH_3^+) in blue; negatively charged groups (COO⁻) in red.

Naming the atoms: Atoms in the mainchain of each residue are denoted N, $C\alpha$, C and O. The sidechain is attached to the $C\alpha$. Sidechain atoms are identified by their chemical symbol, and by successive letters from the Greek alphabet, proceeding out from the $C\alpha$. Thus, the sidechain of methionine has

atoms C β, C γ, S δ, C ε: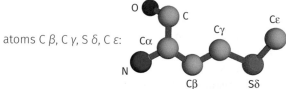

(b) The 20 natural amino acids in space-filling representation. Hydrogen atoms not shown. Grey = carbon, red = nitrogen, blue = oxygen, yellow = sulphur. For glycine and alanine only, all non-hydrogen atoms, including the mainchain atoms N and C=O, are shown; for the others only the C α and sidechain atoms appear.

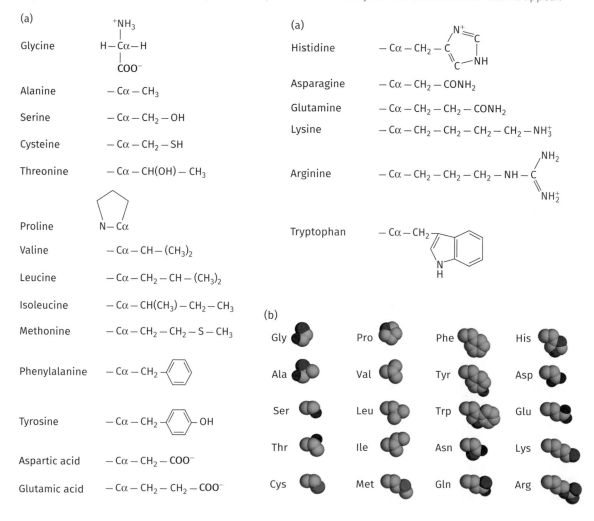

- Some sidechains are electrically neutral: Because of the thermodynamically unfavourable interaction of hydrocarbons with water, residues containing large aliphatic or aromatic neutral sidechains are called 'hydrophobic' residues.

- Some sidechains are polar: Asparagine and glutamine contain amide groups; serine, threonine, and tyrosine, hydroxyl groups. Polar sidechains, and mainchain peptide groups, can form **hydrogen bonds**. (A hydrogen bond is a weak interaction between two electronegative atoms, usually O or N, mediated by a hydrogen between them, for instance: O—HN. Hydrogen bonds vary in strength, but are approximately 1/20 the strength of covalent bonds.)

- Other sidechains are charged: Aspartate and glutamate are negatively charged; lysine and arginine (and, usually, histidine, especially at low pH) are positively charged. The charged atoms occur at or near the ends of the sidechains, which, except for histidine, are relatively long and flexible. The atoms nearest to the backbone are non-polar. Two sidechains with positive and negative charge can approach each other in space to form a 'salt bridge'.

2.3 Protein folding and denaturation

The polypeptide chain has intrinsic flexibility, and could fold up in space into many different conformations. For most proteins, however, there is a unique thermodynamically stable **'native conformation'** that is biologically active. This native conformation depends on the amino-acid sequence. In this way the linear sequence of bases in the genome is translated into a three-dimensional structure (see Figure 1.2).

 Key point

The great variety of three-dimensional structures that proteins can adopt underlies the great variety of their functions.

What stabilizes native states of proteins?

At constant temperature and pressure, chemical systems come to equilibrium at a state of minimal **Gibbs Free Energy**, $G = E + PV - TS$. (Here E = energy, P = pressure, V = volume, T = absolute temperature, and S = entropy.)

The entropy S is a measure of conformational freedom. For instance, if you put a drop of ink in a glass of water, it will spread out throughout the entire volume uniformly. In this process there is no appreciable change in energy or volume and therefore this process is driven entirely by the increase in entropy: by spreading out from confinement in a single drop, to occupying the entire volume, the particles of ink achieve much greater conformational freedom.

For reactions not involving gases, volume changes are small (and we are imposing the condition that pressure must be kept constant). Thus in protein folding PV does not change appreciably. Under these circumstances, the system seeks a compromise between low energy E, and high entropy S.

In the folding of proteins, the denatured state has very high conformational freedom, relative to the native state in which all molecules adopt the same unique conformation. Therefore the folding to the native state involves a great *decrease* in conformational entropy. This severe thermodynamic barrier to forming the native state must be overcome by compensating effects. These include:

1. *The hydrophobic effect*: Sidechains like the benzene ring of phenylalanine have unfavourable interactions with water. In the unfolded, denatured state of a protein, they are exposed to water. A favourable contribution to the protein stability is to bury hydrophobic sidechains inside the compact native structure, out of contact with water. The same reasoning accounts for the phase separation observed in the lab in a water-ether mixture, or in the kitchen in an olive oil-water mixture such as salad dressing.

   Hydrophobicity is an *entropic* effect: Although pure water is a relatively ordered liquid, hydrophobic solutes cause water to be even more ordered around them, reducing conformational freedom and raising the entropy.

2. *Forming hydrogen bonds between atoms of the protein*: A compact native protein structure buries not only hydrophobic sidechains, but polar and charged sidechains, and mainchain nitrogen and oxygen atoms. In the denatured state, they would form hydrogen bonds to water. Depriving them of hydrogen bonds would be an intolerable energetic penalty, which would prevent the formation of the native state. What happens is that hydrogen bonds between water and a protein atom are replaced by hydrogen bonds between protein atoms. These can involve mainchain and/or sidechain atoms.

 Key point

Some sidechains, but not all, can form hydrogen bonds. Referring to Figure 2.3(a), which sidechains can form hydrogen bonds?

3. *Van der Waals forces* occur between all non-bonded atoms; they are the forces that give many ordinary liquids, such as petrol, their cohesion. (In water, instead, hydrogen bonds provide the cohesive forces.) The energy of the system is lower, the closer the atoms are to one another, until they actually collide. The atoms in native states of proteins form as compact a structure as possible, to take advantage of Van der Waals attraction.

Key point

1. When you compress a gas quickly, as in blowing up a bicycle tyre, the molecules approach one another more closely, the attraction between them is greater, and the intermolecular energy goes down. The law of conservation of energy then requires that the the temperature of the system *increases*, heating up the pump, as you will have observed.

2. Water has exceptional properties because of extensive water-water hydrogen bonding in the liquid. The hydrogen bonding explains why water has a much higher boiling point (100°) than methane (–162°), which has almost exactly the same molecular weight.

4. *Additional covalent bonds* sometimes add further stability and rigidity to native structures. The most common of these is the **disulphide bridge**, in which the SH groups of the sidechains of two cysteines that approach each other in the three-dimensional structure react to form a covalent bond:

$$-SH + HS- \rightarrow -S-S$$

The cysteines must be nearby in space, but need *not* necessarily be close together in the amino-acid sequence. The folding of the chain can bring into spatial proximity pairs of cysteines far apart in the sequence.

 Key point

The hydrophobic effect, hydrogen bond formation, Van der Waals forces, and additional covalent bonds such as disulphide bridges, contribute to the stabilization of native protein structures.

The native structures of proteins are determined by their sequences of amino acids

How do we know this? Native structures of proteins are stable under only a limited range of conditions. If heated, or subjected to specific chemicals, the native structure breaks up, or 'denatures'. Differences between denatured and native states include:

1. In most cases, in the native state each molecule adopts a single conformation.
   - In the denatured state each molecule has a different conformation.
2. In the native state the structure is usually dynamically stable, or, in most cases undergoes only small, discrete conformational changes.
   - In the denatured state, the conformations of the molecules are changing, adopting successively a random series of very different conformations.

3. Whereas the native state adopts a compact structure,

— denatured proteins are unfolded to occupy a much larger volume.

Thus, after denaturation the protein has lost all 'memory' of what the native state is. To reform the native state, the only clue available to guide it is the amino acid sequence.

Scientific approach 2.1
Reversible denaturation of proteins shows that amino-acid sequence determines protein structure

In the 1950s, a set of classic experiments by C.B. Anfinsen showed that proteins could exhibit *reversible* denaturation. He used the small protein Ribonuclease A (142 amino acids), an enzyme that hydrolyses RNA. Anfinsen showed that exposure of a dilute solution of Ribonuclease A to guanidinium hydrochloride (which breaks hydrogen bonds between protein atoms) and β–mercaptoethanol (which breaks disulphide bridges) denatured the protein, and caused it to lose enzymatic activity. But upon removing the guanidinium hydrochloride and reoxidizing the disulphide bridges, the protein refolded to the native state, regaining enzymatic activity.

This result has been extended to many other proteins, establishing the *very* important principle: **protein structure is determined by amino-acid sequence**. This deserves an extension to Francis Crick's Central Dogma:

DNA → RNA → amino-acid sequences of proteins → protein structures → protein functions

 Pause for thought

1. Anfinsen's experiments established the general principle that protein denaturation is *reversible*: if a denatured protein is brought back to normal conditions of solvent and temperature it will recover its native structure and function. But as readers may well point out, if one boils an egg and subsequently cools it down, one does not recover a raw egg. Does this contradict Anfinsen's principle?

The difference is that the proteins in an egg are very concentrated. Heating denatures them, and the denatured proteins aggregate. The **aggregation**, not the denaturation, is the irreversible step.

Key point

Old molecular biologist's joke: 'How might you reverse the boiling of an egg? Answer: Feed it to a chicken!'

Anfinsen used dilute solutions, for which denaturation did not lead to aggregation. But are protein concentrations inside cells like the conditions of Anfinsen's experiment, or like an egg? (After all an egg *is* a cell.)

Why, under normal circumstances, do proteins in cells not aggregate? Describe the mechanism whereby certain mutations enhance the risk of intracellular protein aggregation. Describe some of the mechanisms that cells use to protect themselves against intracellular protein aggregation.

2. In some cases, mutations destabilize proteins, sometimes to moderate extents only. α_1-antitrypsin is an elastase inhibitor that protects the lungs from neutrophil elastase. Dysfunction of α_1-antitrypsin raises the risk of emphysema.

The Z-mutant of α_1-antitrypsin, Glu342→Lys, is destabilized to the point that it can denature and aggregate in temperatures attainable in high fevers. Many readers will know that it is not uncommon for ill babies to 'spike' high fevers, with body temperatures above the denaturation temperature for Z-α_1-antitrypsin. Suppose an infant is homozygous for the Z mutant of α_1-antitrypsin. What clinical advice would you give the parents, warning them of what to look out for and what action to take should their infant become ill?

2.4   Cofactors and post-translational modifications

All proteins contain polypeptide chains. In some cases, however, protein structures contain additional substances. *Cofactors* are atomic ions or small organic molecules that form integral and stable parts of proteins.

Cofactors supplement the chemical versatility of the amino acids. Although amino acids have a range of properties and can support many protein functions, there are some things for which they need help. For instance, amino acids are in general not very good at oxidation-reduction. Proteins that participate in **oxidation–reduction reactions** generally contain cofactors such as NADH, which can shuttle between oxidized and reduced forms (see Figure 2.4).

 Key point

Contrast: substrates of enzymes bind to proteins only transiently and are changed during the interaction; cofactors are stable components of the protein structure. Changes, such as the oxidation of NAD (Figure 2.4) are reversible.

Some proteins contain modified amino acids. Disulphide bridges are the most common example. Another important **post–translational modification** is the phosphorylation of sidechains, serine, threonine, and tyrosine, by

Fig. 2.4 Nicotinamide adenine dinucleotide (NAD) is a common cofactor in proteins. It can support oxidation-reduction reactions by interconverting between reduced (NADH, on left) and oxidized forms (NAD$^+$, on right). The two forms have different absorbance at a wavelength of 340 nm; enzymologists exploit this to follow reactions coupled to their interconversion.

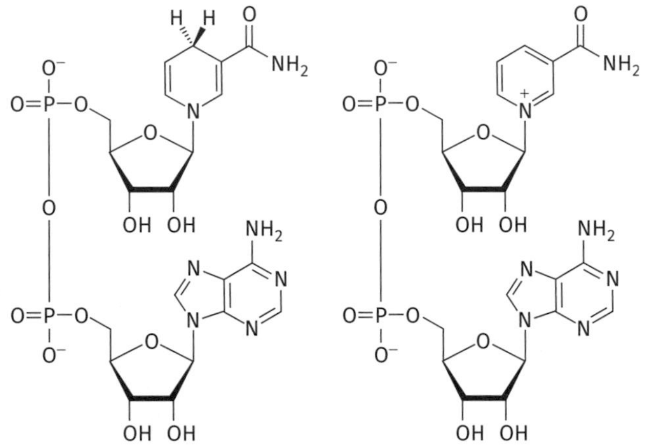

adding a phosphate group to the hydroxyl of the sidechain. Phosphorylation of tyrosine sidechains:

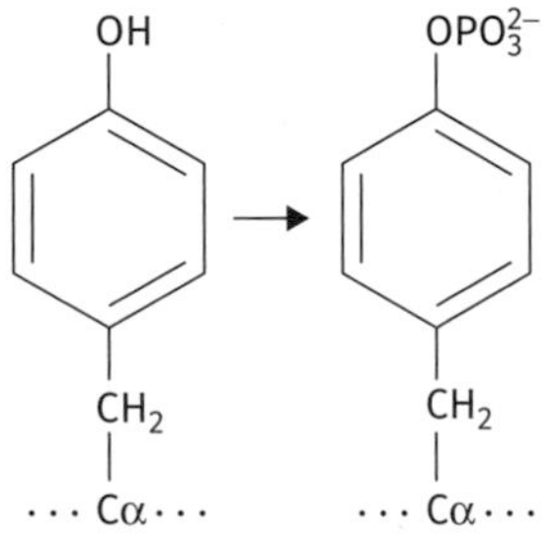

is an important mechanism of activating signal-receptor proteins.

2.5   Protein structures and their analysis

Investigation of protein structures, like connoisseurship in art, requires both a sensitive and trained eye, and technical scientific analysis. The polypeptide chains of proteins in their native states describe graceful curves in space. These are best appreciated by temporarily ignoring the detailed interatomic interactions of sidechains and focussing on the calligraphy of the mainchain. This reveals that the major differences among protein structures come not at the level of local interactions, such as formation of α–helices (see below); in most proteins these substructures are quite similar. The differences appear rather at the level of the three-dimensional assembly of regions, in which the same local substructures are differently deployed in space to give different protein folding patterns.

Figure 2.5 shows the chain tracing of two proteins. The 'chevrons' indicate the direction of the strands. The reader is urged to trace the chains

Fig. 2.5 Chain tracings of two small protein domains. (a) Uteroglobin. (b) Ribosomal protein L7/L12.

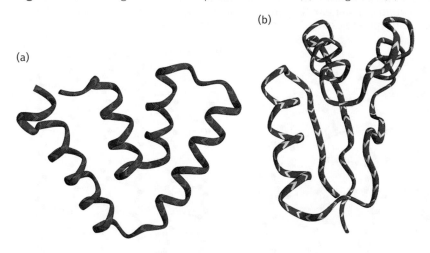

(a)

(b)

visually, from N-terminus to C-terminus. (Example (a) is simpler than part (b).) The rotating versions, see online resources associated with this book, will be useful.

> ### 💡 Key point
>
> Study of protein folding patterns in terms of the curves that the mainchain traces out in space can (1) expose recurring structural patterns, (including but not limited to α–helices and β–sheets), and show the variety of their patterns of combination, (2) provide an approach to comparison of different folding patterns, to clarify structure-function relationships and evolutionary pathways.

Secondary structures: α–helices and β–sheets appear in many proteins

Many proteins show recurrent substructures: Figure 2.6 shows the two most common examples, the α–helix and β–sheet:

α–helices and β–sheets are both stabilized by hydrogen bonds between mainchain atoms. Therefore they are compatible with different sequences of sidechains. α–helices are formed from a single consecutive set of residues in the amino acid sequence (Figure 2.6(a)). They are thus a *local* structure of the polypeptide chain; that is, they form from a set of contiguous residues in the sequence. The hydrogen-bonding pattern links the C=O group of residue i to the H—N group of residue $i + 4$. (From N-terminus to C-terminus defines the direction of a polypeptide chain. Thus residue $i + 4$ is further from the N-teminus of the chain than residue i.) Other types of helices, with different hydrogen-bonding patterns, are rare.

Fig. 2.6 α–helices and β–sheets are two standard structures that appear in many proteins. In this figure, broken lines represent hydrogen bonds. (a) α–helix. (b) parallel β–sheet. (Hydrogen atoms appear in this picture.) (c) antiparallel β–sheet. Can you see examples of these in the structures in Figure 2.5?

(a)   (b)   (c)

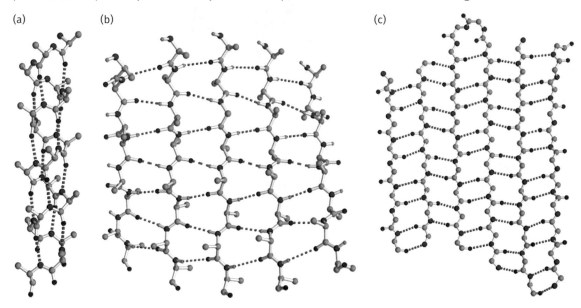

β–sheets form by the lateral interactions of several different local regions of the chain (Figures 2.6(b) and (c)). They can bring together residues separated widely in the amino acid sequence. Figure 2.6(b) shows an ideal β–sheet with all strands parallel. (That is, in all five strands appearing in Figure 2.6(b) the N → C direction is *down* the page. You can see that at the bottom of each strand there is a carbonyl group: .) Each of the three central strands has two neighbouring strands in the sheet. The two edge strands have only one neighbour in the sheet. The strands are not quite fully extended, but accordion-pleated. The sheet is not flat, but each strand is rotated from its neighbours so that the sheet appears twisted in propellor fashion. (This is difficult to see in the 'still' picture in the book, but easily visible in the rotating version, see online resources.)

Figure 2.6(c) shows an antiparallel β–sheet, in which the strands alternate in direction. (For the leftmost strand in Figure 2.6c) the N → C direction is *down* the page. In the neighbouring strand the N → C direction is *up* the page.) β–sheets can be parallel, antiparallel, or mixed, with respect to the relative directions of the strands. In Figure 2.6(c), the central pair of adjacent strands is connected through a short loop, a structure called a β–hairpin. (What are the relative directions of the strands in acylphosphatase, Figure 1.3?)

> 💡 **Key point**

α–helices and β–sheets are two standard substructures that appear in many proteins.

Conformational angles define protein conformations

Angles of internal rotation around successive mainchain atoms define conformations of individual residues.

In proteins, the mainchain in each residue has two single bonds: N–Cα and Cα—C. In principle there is free rotation around these bonds; limited, however, by steric interactions involving sidechain atoms. The third mainchain bond, the peptide bond, has partial double bond character and appears almost always in the *trans* conformation, occasionally *cis*.

Figure 2.7(a) shows the definitions of angles of rotation ϕ, ψ, and ω respectively, around bonds N–Cα, Cα—C, and C–(N of next residue). (The C–N bond is the peptide bond.) A graph showing the angles ϕ and ψ for the residues is a **Sasisekharan-Ramakrishnan-Ramachandran plot** (often shortened to Ramachandran plot). Figure 2.7(b) contains such a plot for acylphosphatase, the molecule shown in Figure 1.3.

Primary, secondary, tertiary, and quaternary structures

Classically, there have been four levels of description of protein conformations. The Danish protein chemist K.U. Linderstrøm-Lang defined the primary, secondary, and tertiary structures. For proteins composed of more than one

Fig. 2.7 (a) Definition of mainchain conformational angles: ϕ = rotation around N—Cα bond; ψ = rotation around Cα—C bond, and ω = rotation around the peptide bond. Verify that in this picture, the peptide bond is *trans*.

(b) Sasisekharan-Ramakrishnan-Ramachandran plot, a graph of ψ and ϕ angles for the residues in acylphosphatase (see Figure 1.3). Sterically most-favourable regions for *trans* peptides in green; sterically allowed regions for *trans* peptides in yellow. Residues with $\phi > 0$, mostly glycines, are identified in red. Because glycine has only a hydrogen for a sidechain, it is subject to less-restrictive steric constraints than other residues.

The two sterically-favourable regions, shown in green, correspond to the conformations adopted by residues in right-handed α–helices (α_R) and β–sheets (β). It is not a coincidence that these structures *both* have favourable mainchain conformations, *and* can form stabilizing hydrogen bonds.

(Changing the signs of ϕ and ψ gives the left-handed α–helix (α_L). Left-handed α–helices are rare in proteins because of steric clashes between sidechains other than glycine.)

(a)

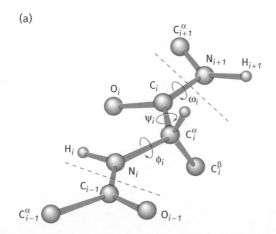

Fig. 2.7 (*Continued*)

(b)

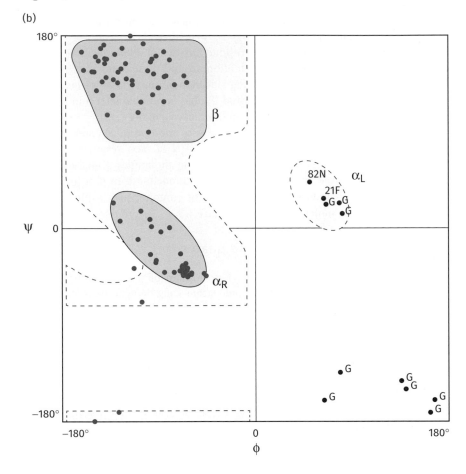

subunit, J.D. Bernal added the level of quaternary structure, how the protein is built of monomers (= individual polypeptide chains):

Primary structure	The covalent bonds that correspond to the amino-acid sequence of the protein.
Secondary structure	The distribution along the polypeptide chain of α–helices and strands of β–sheet.
Tertiary structure	The three-dimensional spatial conformation of the polypeptide chain.
Quaternary structure	For proteins containing multiple polypeptide chains, the composition in terms of monomeric subunits, and the spatial assembly of the monomers.

For instance, for acylphosphatase, the protein illustrated in Figure 1.3:

- The primary structure is the amino acid sequence:

  AEGDTLISVDYEIFGKVQGVFFRKYTQAEG

  KKLGLVGWVQNTDQGTVQGQLQGPASKVRH

  MQEWLETKGSPKSHIDRASFHNEKVIVKLD

  YTDFQIVK

- The secondary structure is the set of assignments:

Type of secondary structure	Residue ranges
strand of β–sheet	6–16
α–helix	22–32
strand of β–sheet	36–41
strand of β–sheet	47–54
strand of β–sheet	55–67
α–helix	55–67
strand of β–sheet	94–97

 Key point

Readers should be able to trace the chain in Figure 1.3(c) visually, to verify that the elements of secondary structure appear in the order indicated in this table.

- The tertiary structure contains two helices packed against a 5-stranded β–sheet. Both helices are on the same side of the β–sheet; their axes are approximately parallel to the strands. Four of the five strands of the β–sheet are antiparallel; the fifth (on the right in the orientation depicted in Figure 1.3(c)) is parallel to its neighbouring strand.
- The protein is monomeric so there is no quaternary structure.

Supersecondary structures

In addition to the common secondary structures, α–helices and β–sheets, many proteins show common patterns of interaction between sets of α–helices and strands of β–sheet that appear (a) consecutively in the sequence, and (b) close together in the three-dimensional the structure. Supersecondary structures are local structures, formed by residues in a contiguous segment of the sequence, that contain two or more α–helices and/or strands of β–sheet, packed against each other to form an internal local compact unit. Common supersecondary structures include the α–helix hairpin, the β–hairpin, and the β–α–β unit (see Figure 2.8). Supersecondary structures appear, in the hierarchy of protein architecture, between secondary and tertiary structures. Such small structural patterns that recur in many proteins are also known as motifs.

A picture gallery

Figures 2.5 and 2.9 contain a selection of small protein structures. These pictures illustrate the structural themes that we have discussed. Appreciation of the form, and assembly of the components, of these structures that will gained by studying and analysing the pictures 'by eye' is well worth the effort. The reader is urged to trace

Fig. 2.8 Common supersecondary structures: (a) α–helix hairpin. (b) a β–hairpin. (c) a β–α–β unit. (d) A combination of individual supersecondary structures forms a complete structure. Spinach glycolate oxidase is an example of a very common β–barrel structure (a β–α barrel) formed from a succession of β–α–β units. The eight-stranded sheet is closed into a cylinder. Helices are shown in orange, with yellow chevrons. Alternate strands in the β–barrel are shown in red and blue, with yellow chevrons.

(a)   (b)   (c)

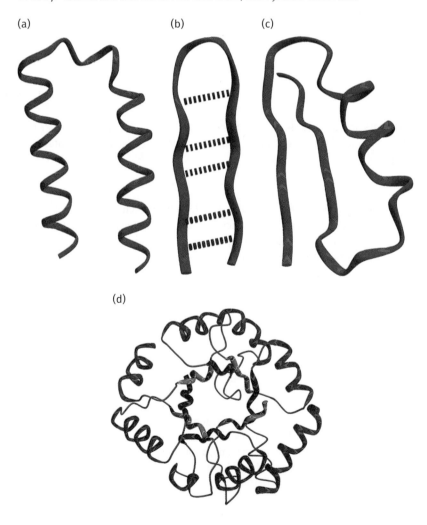

(d)

Fig. 2.9 An album of small protein domains, which will repay study. 'To see is itself a creative operation, requiring an effort.' H. Matisse. (a) Engrailed homeodomain. (b) Phospholipase A$_2$. (c) T4 Lysozyme. (d) Timothy grass pollen allergen. (e) Abl tyrosine kinase/peptide complex (two views). (f) Bovine pancreatic trypsin inhibitor. (g) Ribonuclease T1. (h) Barley chymotrypsin inhibitor. (i) Malate dehydrogenase, NAD-binding domain. (j) Alanine racemase, N-terminal domain. (k) Wheat-germ agglutinin.

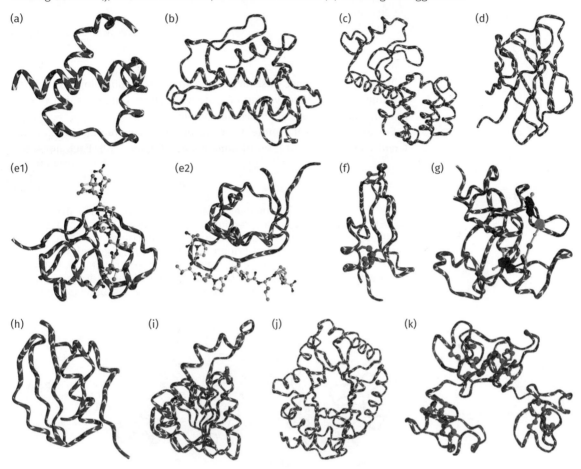

the chains visually, picking out the helices and sheets, and disulphide bridges (are the residues they link close or distant in the sequence?) Can you see supersecondary structures? Consider these pictures as exercises in training your eye to recognize the important spatial patterns. What is unusual about wheat germ agglutinin?

Domains

A *domain* is a compact substructure within a polypeptide chain, that appears to have independent stability. A domain is intermediate between secondary and tertiary structure, but larger than a supersecondary structure. Many proteins contain several domains; these are called modular proteins (see Figure 1.7). It is not uncommon for a set of independent monomeric proteins in prokaryotes to combine to form a corresponding multidomain protein in eukaryotes. (The converse also occurs—a multidomain protein in prokaryotes corresponding to separate monomeric proteins in eukaryotes—but more rarely.)

💡 **Key point**

Distinguish *domains*—independent structural units within a single poly-peptide chain—from *multimeric proteins*—assemblies of separate polypep-tide chains.

To follow divergence of sequence and structure in homologous proteins, the domain is the effective unit of protein structure. Databases that classify protein conformations are based on domains.

For example, several proteins active in blood coagulation arise as dif-ferent combinations of a set of domains (see Figure 2.10). Each domain is

Fig. 2.10 Several proteins involved in blood coagulation show structures based on the same domains, recombined in different ways. The figure illustrates the structures of six domains, and an icon symbolizing them: SerPr = serine protease, Kr = kringle, EGF = epidermal growth factor, Fn2 = fibronectin 2, Gla = γ–carboxyglutamic acid-rich domain, Fn1 = fibronectin 1. The figure shows choice and order of these domains in the proteins: Factor XII, Factor IX, Prothrombin, and Tissue Plasminogen Activator.

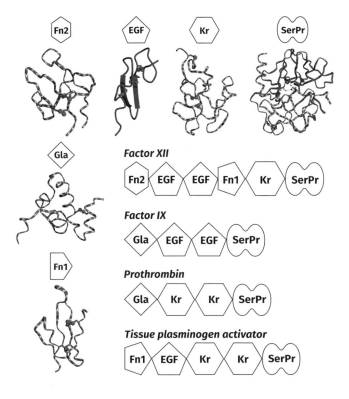

a relatively small compact unit in its own right. The domain composition and the order of domains along the polypeptide chain differ among these proteins.

Key point

Different proteins can 'mix and match' sets of domains, thereby creating novel structures and functions without having to invent new folding patterns.

Conformational change

Crystal structures portray protein structures in fixed native conformations. The constraint of packing into a periodic crystal structure almost always *forces* the molecules to adopt a unique conformation. This made for a very simple and attractive picture: The fully-disordered denatured state v. the native state with a unique conformation. And yet there was evidence that this picture was misleading. There was suspicion that crystal structures might be statues, capturing only one pose of a dynamic and active subject.

Despite the tenet of the field that proteins adopt unique native states, many proteins undergo conformational changes as part of the mechanisms of their normal activities. In many but not all cases binding of ligands triggers the change, called induced fit (see Figure 2.11). In most cases the different conformations retain the same folding pattern, but there are exceptions in which the topology of the tertiary structure changes. Proteins can also give up their native conformations to adopt a characteristic β–sheet structure, in amyloid fibrils.

Key point

Many proteins undergo conformational changes as part of their normal function.

Types of conformational changes include:

* *Open and closed forms*: The enzyme phosphoglycerate kinase binds substrate and cofactor in a cleft between two domains. In the un-ligated form, the two domains swing open to receive the ligands. In the ligated form, the domains close over the ligands, excluding water from the active site (see Figure 2.11). Such 'hinge motions' are a common mechanism of conformational change: The two domains remain individually rigid but change their relative orientation, by means of structural changes in only a few residues in the regions linking the domains. Another example: Hinge motion in myosin is responsible for the mechanical impulse in muscle contraction.

Fig. 2.11 Phosphoglycerate kinase catalyses the reaction in glycolysis:

1,3-bisphosphoglycerate + ADP ↔ 3-phosphoglycerate + ATP

(abbreviations: 1,3-bisphosphoglycerate, 1,3BPG; ADP, adenosine diphosphate; 3-phosphoglycerate, 3PG; ATP, Adenosine triphosphate)

Phosphoglycerate kinase contains an N-terminal domain (green) which binds 1,3-bisphosphoglycerate (1,3BPG) and a C-terminal domain (red) which binds ADP. The links between the domains, shown in blue, act as a hinge. On the left, the enzyme is 'open' to allow binding of both substrates, but they are not in the proximity required for the reaction. A rotation around a hinge region achieves the 'closed form' which brings the substrates together (centre), and the reaction occurs. In the centre image the phosphate group being transferred is shown in grey. The right image shows the enzyme bound to the products, 3-phosphoglycerate (3PG) and ATP.

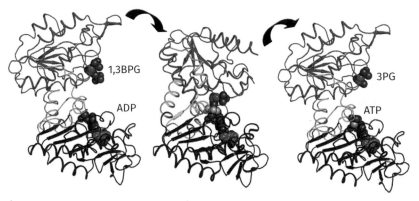

(Reprinted from FEBS letters, 587/13, Matthew W Bowler, Conformational dynamics in phosphoglycerate kinase, an open and shut case?, Page 1879, Copyright (2013), with permission from John Wiley & Sons.)

 Key point

The reaction catalysed by phosphoglycerate kinase shows a *direct* coupling of a reaction of glycolysis to synthesis of ATP. The electron-transport chain achieves ATP synthesis by a much more complex mechanism (see Chapter 4).

- *Allosteric proteins* show another class of conformational changes induced by ligand binding. Various types of conformational changes are observed. Many allosteric changes involve changes in quaternary structure.
  - For instance, haemoglobin, an $\alpha_2\beta_2$ tetramer, shows an allosteric change involving a rotation of about 15° of one $\alpha\beta$ dimer with respect to the other. Changes in tertiary structure within the subunits are coupled to the quaternary structure change.

- *ATP synthase* shows a very large change—it is a rotatory motor in which the turning of an 'axle' induces successive conformational changes in the subunits of a hexamer. These conformational changes drive the synthesis of ATP.

- *GroEL–GroES* is a 'chaperone' complex that rescues misfolded proteins. The complex is a large reciprocating engine that depends for its activity on a crucial conformational change in the GroEL protein. This conformational change is coupled to the unfolding of misfolded proteins, which—like juvenile offenders—are released to have another chance to shape up (i.e. fold properly).

- *Localized conformational changes in loop regions*. Loops—regions between successive secondary-structural elements—are generally the most flexible regions in proteins. In enzymes such as triose phosphate isomerase and p21 ras, a change in ligand binding affects the conformation of local mobile regions.

- *Icosahedral viruses*. Most viruses assemble capsids from multiple copies of a single coat protein. In many cases different coat proteins have the same amino-acid sequence but adopt different conformations, in order to satisfy the symmetry requirements of the polyhedral array.

- *A few proteins appear to have more than one possible folding pattern, with different topologies*. These include the prion protein and the serpins (a family of serine protease inhibitors). The conformational change in prion proteins critically affects their infectivity. In the case of serpins, the conformational change not only inhibits, but denatures, the target protease.

- *Amyloid aggregates*. Many proteins can form fibrillar aggregates with structures different from their free native state. These are associated with several diseases (see section 2.8).

Intrinsically-disordered proteins (IDP)

Many proteins that exhibit conformational changes alternate between two (or sometimes more) *discrete* states. Some proteins can deviate even further from the idea of a unique and rigid native state, by containing regions that are intrinsically disordered, *even* in the native state. In some but not all cases, the unliganded state shows **intrinsic disorder**, and binding a ligand rigidifies it.

Indeed, binding alternative partners may rigidify a disordered protein to two different structures. The nuclear coactivator binding domain (NCBD)—also called the IRF-3 binding domain (IBiD)—of the creb-binding protein (CBP) is disordered in the unligated state. (The creb-binding protein has the function of activating transcription.) NCBD forms two distinct proteins by binding to alternative partners: ACTR domain of p160 (which regulates transcription factors), and Interferon Regulatory Factor 3. The N-terminal domain of NCBD adopts different conformations in complex with the two partners.

A still further generalization is a 'fuzzy complex' in which even the conformation of the ligated state is not rigid. Such a complex may pay a price—in terms of affinity—relative to a rigid complex, but for functional reasons may require a molecule capable of less-specific and weaker binding.

> 💡 **Key point**
>
> The general principle is that proteins fold to compact native states with a structure dictated by their amino-acid sequences. However, there are important exceptions: some proteins undergo conformational changes as part of their function; other proteins contain disordered regions, or may even be entirely disordered.

2.6   Mutations

Mutations that alter amino-acid sequences of proteins permit exploration of sequence space.

Many proteins are highly-adapted and finely-tuned for their functions. It follows that many mutations are expected to be deleterious. Indeed, many diseases arise from mutations that impair or even completely prevent normal protein function. Sickle-cell disease, phenylketonuria, and cystic fibrosis are well-known examples. Mutations in BRCA1, BRCA2, and PALB2 genes increase the risk of breast and ovarian cancer; many women are testing these sequences in their genomes. The database Online Mendelian Inheritance in Man (https://www.omim.org) collects mutations with clinical consequences.

Mutations are also an important engine of evolution. They allow proteins to modify their functions, and even to develop new ones. We shall discuss protein evolution in Chapter 5.

2.7   Protein families

The haemoglobins of humans and other species share similar but usually not identical amino-acid sequences. They fold into very similar structures and have the common function of transporting oxygen and carbon dioxide in the bloodstream. They are an example of a homologous family of proteins. (Homologous proteins are proteins that have diverged from a common ancestor; see Bigger picture 2.1). Myoglobin, a more-distantly diverged—but still homologous—protein, accepts oxygen from haemoglobin in muscles.

> 💡 **Key point**
>
> Proteins evolve by changes in DNA sequences that give rise to changes in amino-acid sequences, which in turn can alter protein structure and function. Selection acts on protein function, to alter gene frequencies in populations.

Figure 2.12 shows the sequence similarity of several typical globins. Looking only at the three α chains in this assembly—from human, horse, and turkey—it is apparent that the divergence of the sequences mirrors the

The bigger picture 2.1
The distinction between similarity and homology

- Similarity is the observation or measurement of resemblance and difference, independent of the origin of the resemblance.
- Homology means, specifically, that the sequences and the organisms in which they occur are descended from a common ancestor. The implication is that similarities derive from shared ancestral characteristics.
- Similarity of sequences or structures is observable *now*, and involves no historical hypotheses.
- In contrast, assertions of **homology** are statements about historical events that are almost always unobservable. *Homology must be inferred from observations of similarity.*
- Only in a few special cases is homology directly observable; for instance in pedigrees of human families carrying Huntington's disease; or in laboratory populations of animals, plants, or microorganisms; or in clinical studies that follow sequence changes in viral infections in individual patients.
- For proteins that have diverged very widely, with only tenuous similarities appearing in the amino-acid sequences and the structures, it can be difficult to decide whether they are homologues or not.

Discussion Questions

1. Suppose that in a multiple sequence alignment of a family of proteins, one position contains alanine in 50% of the sequences and phenylalanine in 50% of the sequences. What other amino acids would you expect might appear at that position in other homologous structures, consistent with retaining the structure and function of the protein?

2. For all the sequences shown in Figure 2.12, there is enough similarity to justify the conclusion that they proteins are homologues. But consider the following:

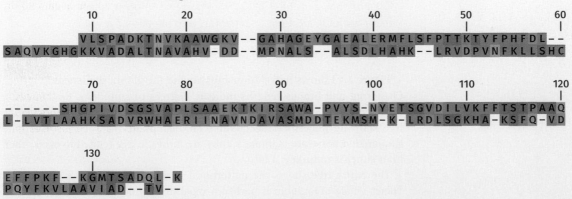

On the basis of this aligned pair of sequences, could you confidently decide whether or not they are homologues? How could you test the hypothesis that they are homologues? What types of evidence would you want to adduce?

Fig. 2.12 An **alignment** of the amino-acid sequences of the α and β subunits of human and horse haemoglobin (Hb), the α subunit of turkey haemoglobin, and sperm whale myoglobin (Mb). Different colours indicate different physico-chemical types of amino-acid sidechains: green = large hydrophobic; yellow = small hydrophobic; pink = polar, uncharged; red = negatively charged; and blue = positively charged. Upper-case letters in black below each band indicate residues conserved in all sequences. Lower-case letters in black below each band indicate residues conserved in all but one sequence.

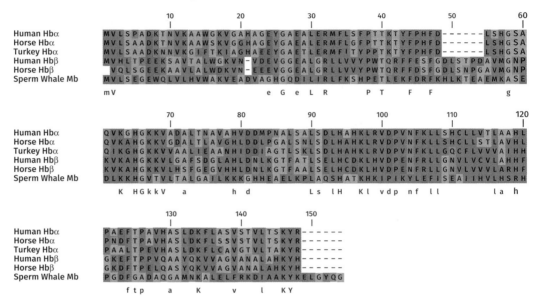

divergence of the species. Indeed, often DNA and amino-acid sequence data are the most reliable basis for establishing phylogenetic relationships among species.

 Key point

The reader should verify, by counting identical corresponding residues that the most similar pairs of sequences are the α subunits of human and horse haemoglobin, and the β subunits of human and horse haemoglobin; and that myoglobin has fewest similarities with the other five.

Figure 2.13 shows the relationships among the globin sequences as a tree. Clustering and branch lengths indicate degrees of similarity. Restricted to the haemoglobin α chains, the tree has the same structure as a phylogenetic tree showing ancestor-descendant relationships of the species of origin: Human and horse are mammals; they are more closely related to each other than either is to turkey, a bird.

The entire tree shows the patterns of divergences of all six molecules. Sperm whale myoglobin is part of a separate globin subfamily. It does not map, in the phylogenetic tree in Figure 2.13, where whale might be expected to, from the taxonomic relationship of the species.

Fig. 2.13 Tree diagram showing clustering of similar sequences (HHbα = human haemoglobin, α chain, EHbα = horse haemoglobin, α chain, THbα = turkey haemoglobin, α chain, HHbβ = human haemoglobin, β chain, EHbβ = horse haemoglobin, β chain, SWMb = sperm whale myoglobin.) Short branches indicate similar sequences—for instance the human and horse haemoglobin α chains. Longer branches indicate more-dissimilar sequences; thus sperm whale myoglobin is the outlier.

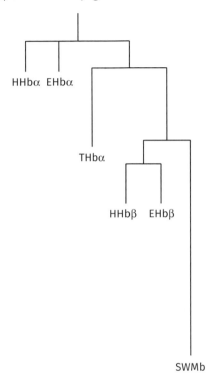

> ### 💡 Key point
>
> Sketch into the tree in Figure 2.13 the place where you would expect the α chain of sperm whale haemoglobin to appear.

The globins are an example of the many families of proteins known. The databases SCOP (scop.mrc-lmb.cam.ac.uk) and CATH (https://www.cathdb. info) collect and present structural relationships among domains.

Families play an important role in elucidating protein structure, function, and evolution. Multiple sequence alignments, such as the one shown in Figure 2.12, reveal patterns of conservation, which can help identify functionally important residues. For instance, the globins in Figure 2.12 contain two conserved histidine residues, at positions 65 and 94 in the alignments. (These correspond to residues 58 and 87 in the sequence of human haemoglobin, α–chain.) These are in the binding site for haem, and interact with the iron of the haem group.

Case study 2.1
The tryptophan synthase pathway

The *trp* operon in *E. coli* begins with a control region containing promoter, operator, and leader sequences.

Control region

◄	trpE	trpD	trpC	trpB	trpA

Five genes encode proteins that catalyse successive steps in the synthesis of the amino acid tryptophan from its precursor chorismate:

Chorismate → anthranilate → phosphoribosylanthranilate →
        (1)   (2)   (3)

indoleglycerolphosphate → indole → tryptophan
              (4)   (5)

- Reaction step (1): *trpE* and *trpD* encode two components of anthranilate synthase. This tetrameric enzyme, comprising two copies of each subunit, catalyses the conversion of chorismate to anthranilate.

- Reaction step (2): the protein encoded by *trpD* also catalyses the subsequent phosphoribosylation of anthranilate.

- Reaction step (3): *trpC* encodes another bifunctional enzyme, phosphoribosylanthranilate isomerase–indoleglycerolphosphate synthase. It converts phosphoribosylanthranilate to indoleglycerolphosphate, through the intermediate, carboxyphenylaminodeoxyribulose phosphate.

- Reaction steps (4) and (5): the next enzyme, tryptophan synthase, catalyses *two* reactions in the pathway: indole-glycerol-phosphate →indole → tryptophan:

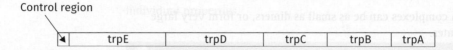

indole-3-glycerol phosphate   indole   tryptophan

- The genes *trpB* and *trpA* encode the β and α subunits, respectively, of a third bifunctional enzyme, tryptophan synthase. The enzyme is a tetramer, containing two α and two β subunits. The α subunits catalyse the first step, formation of indole and glyceraldehyde-3-phosphate. The β subunits catalyse the second step, formation of tryptophan from indole. An allosteric change coordinates the activities of the subunits.

- Why are the two catalytic activities linked in the same protein? The active site of each α subunit is connected to an active site in a β subunit by a 25 Å-long hydrophobic channel contained entirely *within* the enzyme. This conveys the intermediate, indole, from the α to the β active site *without exposure to water*. Exposure to water would permit a side reaction of water with an intermediate in the reaction at the β site, α–aminoacrylate.

Discussion Questions

1. In the 1940s, G. Beadle and E. Tatum observed that single mutations could knock out individual steps in biochemical pathways. On this basis they proposed the **one gene, one enzyme hypothesis**. On a photocopy of the diagrams in this Case study, draw lines linking genes to numbered steps in the sequence of reactions in the pathway. To what extent do the genes of the *trp* operon satisfy the one gene, one enzyme hypothesis, and to what extent do they present exceptions?
2. There is a mapping of the order of appearance of the genes in the DNA and the order of the steps in the metabolic pathway that they catalyse. How might you explain this?

Diseases of protein aggregation

The ability of proteins to form complexes is essential for life. But when this propensity escapes control it can be harmful, even fatal.

The first condition attributed to protein aggregation was sickle-cell disease. As a result of a single point mutation, the deoxy form of haemoglobin forms polymers that precipitate within the red blood cell.

Many diseases are now recognized to arise from protein aggregation (see Table 2.1). Mutations can aggravate the problem—mutated proteins are more prone to misfold, and misfolded proteins are more prone to aggregate. Overproduction of proteins as a result of breakdown of control mechanisms, as in myelomas that overproduce immunoglobulin light chains, also increases the threat of aggregation.

 Key point

Protein aggregation is responsible for many important diseases.

Table 2.1 Diseases associated with protein aggregates

Disease	Aggregating protein	Comment
Sickle-cell disease	deoxyhaemoglobin-S	mutation creates hydrophobic patch on surface
Classical amyloidoses	immunoglobulin light chains, transthyretin, many others	extracellular fibrillar deposits
Emphysema associated with Z-antitrypsin	mutant α_1–antitrypsin	destabilization of structure facilitates aggregation
Huntington's disease	altered huntingtin, with expanded polyglutamine tracts	one of several polyglutamine repeat diseases
Parkinson's disease	α–synuclein	found in Lewy bodies
Alzheimer's disease	$A\beta$, τ	$A\beta$ 40–43 residue fragment of Amyloid Precursor Protein
Spongiform encephalopathies	prion proteins	infectious, despite containing no nucleic acid

Alzheimer's disease

Alzheimer's disease is a neurodegenerative disease common in older people. It is associated with two types of deposits:

1. dense insoluble extracellular protein deposits, called senile plaques. These contain the β fragment (the N-terminal 40–43 residues) of a cell surface receptor in neurons, the β–protein precursor (βPP), or amyloid precursor protein (Figure 2.14(a)).

2. neurofibrillary tangles, twisted fibres inside neurons, containing microtubule-associated protein tau. There is some evidence that amyloid deposits promote tangle formation (Figure 2.14(b)).

Amyloidoses

Classical amyloidoses involve extracellular accumulations of insoluble fibrils, typically 80–100 Å in diameter. Common features include: characteristic microscopic appearance after staining with haematoxylin/eosin; regular fibrils seen in electron micrographs; β–sheet structure in X-ray diffraction, with the backbone perpendicular to the fibre axis: the cross–β structure (see Figure 2.14(c)); bright green fluorescence under polarized light after staining with the dye Congo Red, specific for β–sheet structures; solubility at low ionic strength.

 Key point

Although the name amyloid implies polysaccharide, these aggregates contain protein. Virchow, in the mid-nineteenth century, named them amyloid because, like starch, they could be stained with iodine.

Fig. 2.14 (a) Amyloid plaques in an Alzheimer's brain. Length of horizontal black bar, for scale: 125μm. (b) Neurofibrillary tangles. Length of horizontal black bar, for scale: 62.5μm.

(c) Aβ fragments of the amyloid precursor protein form deposits in the brain, associated with Alzheimer's disease. This structure show the cross −β structure of classic amyloid protein aggregates. The fragment forms extended stacks of antiparallel β−sheets.

(a)

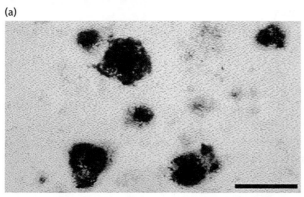

(b)

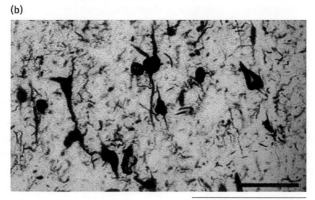

TRENDS in Molecular Medicine

(c)

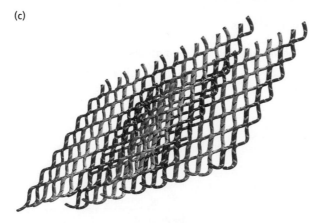

(Parts (a) and (b) Reprinted from Trends in Molecular Medicine, 11/4, Frank M. LaFerla, Salvatore Oddo, Alzheimer's disease: Aβ, tau and synaptic dysfunction, Pages 170–6, Copyright (2005), with permission from Elsevier.
I thank Drs L. Serpell, E.D.T. Atkins, and P. Sikorski for the coordinates of their structure.)

Many proteins are known to form amyloid fibrils, including immunoglobulin light chains and their fragments, transthyretin, and lysozyme. Although these proteins do not have a common folding pattern in their native states, they can apparently adopt a common amyloid structure.

It has been suggested that all proteins can form this common cross–β fibrillar structure, under suitable conditions.

Prion diseases

The prion protein can exist in two forms: the normal PrP^c and the dangerous PrP^{Sc}. PrP^C = prion protein–Cellular; PrP^{Sc} = prion protein–Scrapie. (Scrapie is a prion disease of sheep.) PrP^{Sc} but not PrP^c can:

1. form aggregates
2. catalyse the conversion of additional PrP^c to PrP^{Sc} within the brain of an individual person or animal, and
3. infect other individuals, by various routes, including ingestion of nervous tissue from an affected animal (or person, in the case of kuru).

> ### 💡 Key point
>
> It is a paradox and a puzzle why prions are infective, although not containing nucleic acid.

Prion diseases present widespread health problems for humans and animals (see Table 2.2). In the mid 1990s a serious epidemic of bovine spongiform encephalopathy (BSE) (colloquially, 'mad-cow disease') devastated the United Kingdom countryside. There was an apparent association with the appearance of human cases of variant Creutzfeld–Jacob disease (vCJD).

Table 2.2 Some diseases associated with prion proteins

Disease	Species affected	Symptoms
Scrapie	Sheep	Hypersensitivity, unusual gait, tremor
Bovine spongiform encephalopathy, or 'mad cow disease'	Cow	Similar to scrapie
Kuru	Human	Loss of coordination, dementia
Creutzfeld-Jacob disease (CJD)	Human, age 55–75	Impaired vision and motor control, dementia
variant CJD	Human, age 20–30	Psychiatric and sensory anomalies preceding dementia
Gerstmann-Straüssler-Scheinker syndrome	Human	Dementia
Fatal familial insomnia	Human	Sleep disorder

Whereas in the hereditary disease, familial CJD, symptoms begin to appear in people aged 55–75, variant CJD affects people in their 20s. It is hypothesized that these outbreaks were associated with transmission of prion protein infections across species barriers—sheep to cows for BSE, and cows to humans for vCJD. To stop the outbreak, 4.5 million cattle were culled.

Summary Points

- The principles of construction and design of proteins start with amino acids and build up to complete structures.

- There are 20 standard amino acids. The variety in their properties contributes to proteins' great versatility. Amino acids form peptide bonds, to create a polymer chain. Each amino acid within a polypeptide chain contains a sidechain. The sequences of sidechains—dictated primarily by the DNA sequences of the genes—determine the three-dimensional structures and, thereby, the functions of proteins. In addition to the polypeptide chain, many proteins contain atomic ions and/or small organic molecules; or, some undergo covalent post-translational modifications.

- Most proteins form native states, folding into a compact three-dimensional structure dictated by the amino-acid sequence. In contrast, the denatured state arises when conditions of temperature or solvent break up the native state. Contrast the native state—in which in most cases each protein in a sample takes on the *same* structure—and the denatured state—in which the chain has no preferred structure, and the conformation is mobile. In dilute solution, denatured proteins can recover their native states if brought back to normal conditions.

- In pictures of protein structures showing the tracing of the main-chain, it is possible to recognize common substructures, including α–helices and β–sheets; and supersecondary structures, combinations of α–helices and strands of β–sheets characterizing local regions of the chain.

- Many proteins comprise separate domains—larger subunits that have independent stability; and, indeed, usually show similar folding patterns in different structural contexts arising from assemblies of different combinations of domains.

- Corresponding proteins in different species diverge—to greater or lesser extents—in amino-acid sequence, structure, and function. It is possible to recognize families of related proteins with shared properties; for instance, globins.

- Proteins can interact to form complexes. Many are natural and contribute to normal, healthy function. Others, especially aggregates of denatured or misfolded proteins, can cause disease.

Further Reading

Babu, M.M., Kriwacki, R.W. and Pappu, R.V. (2012). Versatility from protein disorder. Science 337, 1460–1.
The role of flexibility and disorder in protein function.

Berman, H.M., Goodsell, D.R. and Bourne, P.E. (2002). Protein structures: from famine to feast. Am. Sci. 90, 350–9.
General description of the development of protein structure determination and the archiving, curation, and distribution of the results.

Branden, C.-I. and Tooze, J. (1998). Introduction to Protein Structure. 2nd edn (New York: Garland Publishing Co.).
A fine introductory textbook.

Brenner, S. (2003). Nobel lecture. Nature's gift to science. Biosci. Rep. 23, 225–37.
Brenner's profound insights are always worth studying.

Chothia, C. (1984). Principles that determine the structure of proteins. Annu. Rev. Biochem. 53, 537–72.
A classic, still well worth reading.

Janin, J. and Sternberg, M.J.E. (2013). Protein flexibility, not disorder, is intrinsic to molecular recognition. F1000 Biol Rep. 5, 2.
A discussion by two of the leading experts in protein science, of the distinction between flexibility and disorder, and their role in creating ligand-binding sites.

Lees, J.G., Dawson, N.L., Sillitoe, I. and Orengo, C.A. (2016). Functional innovation from changes in protein domains and their combinations. Curr. Opin. Struct. Biol. 38, 44–52.
Discussion of protein evolution at the domain level.

Richards, F.M. (1991). The protein folding problem. Sci. Am. 264(1), 54–7, 60–3.
A general introductory-level article by one of the founders of protein science, responsible for classic work introducing the computational approach to the subject.

 ## Discussion Questions

2.1 Be assigned, or allowed to choose, a multimeric protein of known structure. Present its primary, secondary, tertiary, and quaternary structure. Identify supersecondary structure elements within the structure, and domains. (The web site of the world-wide protein databank, https://www.ebi.ac.uk/pdbe/ will be an essential tool for addressing this question.)

2.2 Chymotrypsin and subtilisin are two proteases. Aligning the sequences shows no obvious similarities. However, both have at their catalytic centres the Ser-His-Asp catalytic triad. These proteins have the same function and the same mechanism. Do you think that this suggests that they are homologous? Consider the structures (these diagrams 'zoom in' on the catalytic triads in the two proteins):

(a)   (b)

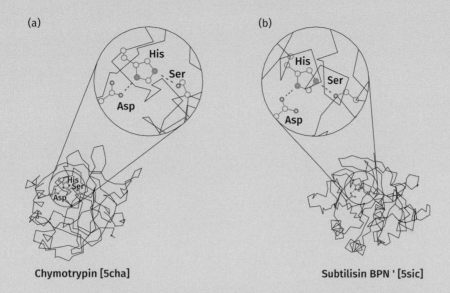

Chymotrypin [5cha]   Subtilisin BPN ' [5sic]

Can you reach any confident conclusions about whether or not the two proteins are homologues, from the structures?

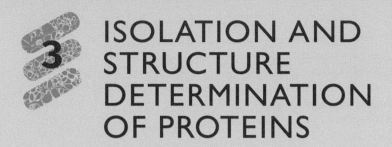

3 ISOLATION AND STRUCTURE DETERMINATION OF PROTEINS

Learning Objectives

- Recognize that in preparing a pure protein from a natural source, we are usually presented with a complex mixture, in which the target protein may well be only a minor component. You will be familiar with the tools available for isolating proteins from mixtures.

- Know about the major experimental methods for determining protein structures: X-ray crystallography, NMR spectroscopy, and cryo-electron microscopy. For each experimental method, you will understand the nature of the data produced, and how protein structures are determined from these data.

- Be able to study a picture of a protein structure, and 'parse' it into secondary and supersecondary structures. Be able to study pictures of two related proteins and identify points of similarity and divergence.

- Understand the major methods for protein structure prediction, and the quality of their results. You will imagine the profound impact a truly general and reliable method of structure prediction will have on investigations and applications of proteins.

To study an individual protein experimentally, you must first prepare a purified sample. It is more difficult to study the properties of proteins in mixtures—including but not limited to cellular interiors! Biochemists have developed a toolkit for isolating proteins from mixtures. Once purified, **X-ray crystallography**, **NMR spectroscopy**, and **cryo-electron microscopy** allow determination of protein structures.

Experimental structure determination is much more labour-intensive than determination of amino-acid sequences, now inferred from genome sequences. Hence we know the sequences of very many more proteins than we know the corresponding structures. But, given that amino-acid sequence determines protein structure, it should in principle be possible to write a computer program that predicts protein structures from their amino-acid sequences. This would unleash all the three-dimensional information implicit in the databanks of genome and amino-acid sequences. It would support the design of unnatural proteins with properties that allow useful applications. **Recent breakthroughs in design of algorithms have spectacularly brought this effort to fruition.**

3.1   Protein purification

Classically, the problem of protein isolation involved purifying a protein from a mixture containing many other molecules, such as might arise from homogenizing a sample of bacterial cells or a beef heart. In contemporary practice, one can insert a gene encoding the target protein into a cell— often, *E. coli*—and induce the cell to overexpress the inserted gene.

Suppose however one starts with a heterogeneous mixture. Each step in a protocol to isolate a particular target protein will involve choosing some physical property, and separating the molecules according to the value of that property. This will bring together the target protein and other proteins from the mixture that share the value of the selected property. For example, selecting molecules on the basis of molecular volume will enrich the mixture in the target protein but retain also others that have similar sizes. Isolating the target will usually require several consecutive steps, using, at each step, a different physical property, and a different method of selection (Table 3.1).

Table 3.1 Techniques for purification of proteins from a mixture

Property	Experimental method
Solubility	ammonium sulphate precipitation
Size/Shape	size-exclusion chromatography
Charge	ion-exchange chromatography
Specific binding	affinity chromatography

 Key point

Isolation of a particular protein from a mixture is generally a multistep process, each step segregating components of a mixture with a particular value of a different physical property, for instance size or charge. After several steps the remaining sample will retain only proteins with specified values of many different physical properties. The goal is that the overall procedure produces a sample containing a unique protein. (See footnote in section 1.5, Proteomics).

Ammonium sulphate precipitation

Ammonium sulphate precipitation is a method for purifying proteins based on differences in their variation in solubility as a function of the concentration of dissolved salts.

Protein solubility requires that proteins do NOT interact strongly with one another—protein-protein interactions can lead to aggregation and precipitation. Charges on proteins mediate interactions—protein surfaces bristle with charges, and a negatively-charged sidechain on the surface of one protein will attract a positively-charged sidechain on the surface of another. In very dilute salt solution, this leads to aggregation and precipitation (Figure 3.1(a)).

Conversely, proteins will be more soluble if they interact more strongly with solvent than with one another. In moderately-concentrated salt solutions (Figure 3.1(b)), the ions of the salt cluster around the charged groups

Fig. 3.1 Protein solubility is determined by a competition between protein-protein interactions and protein-solvent interactions. In these figures large discs represent protein molecules, and small discs represent ions arising from dissociation of salt, perhaps NH_4^+ (pink) or SO_4^{2-} (yellow). (a) Very dilute salt concentration. Insufficient concentration of salt ions to shield protein charges; protein molecules attract one another and aggregate. (b) Moderate salt concentration: salt ions screen charges, protein molecules soluble. (c) High salt concentration: water tied up by salt ions, protein molecules dehydrate, aggregate, and precipitate.

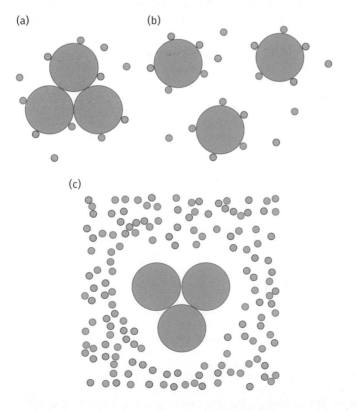

(a)   (b)

(c)

on the protein surface, and screen the electrostatic interactions. As a result, with increasing salt concentration, the solubility of the protein *increases*; this is called salting in.

However, salt ions *compete* with proteins for interaction with water. At high salt concentrations, 'the salt wins'—the protein molecules dehydrate, aggregate, and precipitate; this is called salting out (Figure 3.1(c)).

Figure 3.2 shows the variation of solubility of a protein with the concentrations of different salts. (The ionic strength of a solution is measurement of salt concentration: $I = \frac{1}{2}\sum_i c_i z_i^2$, where c_i is the concentration of the type of ion, and z_i is the charge on the ion.)

In Figure 3.2 the curve corresponding to the salt commonly used in protein purification, ammonium sulphate, $(NH_4)_2SO_4$, is shown in red. Although the general picture is the same for different salts, different ions have different solvation energies and affect protein solubility in individual ways.

The application to protein purification depends on these observations. If the salt concentration in a solution of a protein is increased, typically the protein will precipitate when the salt concentration reaches a threshold. The threshold concentration for precipitation differs between different proteins, making it the basis of a useful technique for protein purification. Traditionally ammonium sulphate is the salt used.

Suppose your target protein will precipitate at an ammonium sulphate concentration of 3 M. Then the basic procedure is to raise the salt concentration in your solution to, perhaps, 2.5 M. Then some proteins will have precipitated but the target has remained soluble. Remove the insoluble material, by centrifugation or filtration, and retain the supernatant (Figure 3.3(a)). Then raise the salt concentration to a value *above* the threshold, perhaps 3.5 M (Figure 3.3(b)), and again separate precipitate and supernatant. This time discard the supernatant (Figure 3.3(c)), and redissolve the precipitate (Figure 3.3(d)). This will produce a new solution enriched in the target protein.

Fig. 3.2 Variation with salt concentration of solubility of haemoglobin, for different salts. Note increase in solubility at low concentration ('salting in') and decrease at high concentration ('salting out'). The difference in the dependence of protein solubility in different concentrations of ammonium sulphate, $(NH_4)_2 SO_4$ for different proteins provides a very popular method for protein purification. (Where might frames (a), (b) and (c) of Figure 3.1 appear in this figure?)

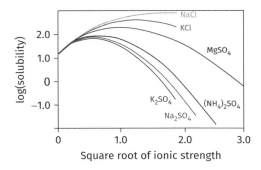

(From: Salis, A. and Ninham, B. (2014). Models and mechanisms of Hofmeister effects in electrolyte solutions, and colloid and protein systems revisited. Chem. Soc. Revs. 43, 7358–77, based on Green, A.A. (1932). The solubility of hemoglobin in solutions of chlorides and sulfates of varying concentrations. J. Biol. Chem. 95, 47–66.)

Fig. 3.3 A mixture of proteins can be enriched in a component targeted for isolation by taking advantage of differential solubility in solutions of different salt concentration. (a) A mixture of soluble proteins at low salt concentration. (b) Addition of increasing amounts of ammonium sulphate reaches a threshold at which the target protein (blue) precipitates. (Some but not all other proteins remain in solution.) (c) centrifuging followed by pouring off the supernatant, leaves the precipitate in a 'pellet', which can be (d) redissolved to form a solution enriched in the target protein.

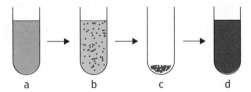

Size-exclusion chromatography

It is possible to create beads with cavities of uniform volume. Packing a chromatography column with these beads, and running a mixture through it, will result in a separation of molecules with volumes *smaller* than the cavity, from molecules with volumes *larger* than the cavity (see Figure 3.4(a)). Large molecules will not enter the crevices in the gel; they will pass unimpeded through the column. Small molecules will take 'detours' into the crevices, and their passage through the column will be retarded (Figure 3.4(b)). Collection of successive fractions from the column will allow separation of larger from smaller molecules.

Fig. 3.4 (a) A polysaccharide bead (purple) contains a crevice. Small molecules (pink) can fit into the crevice; large molecules (blue) cannot. (b) If a column is packed with such beads, running a mixture of different-sized molecules through it will result in retardation of the passage of smaller molecules, and unimpeded passage of the large molecules. Therefore the 'blue' molecules will emerge from the column before the 'red' molecules, achieving a separation on the basis of molecular size.

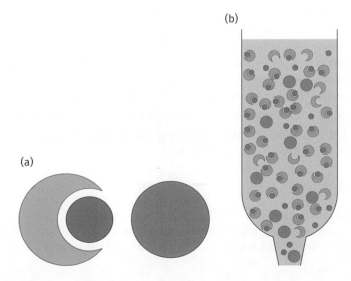

Because inorganic salts are much smaller than proteins, size-exclusion chromatography is a very convenient way to 'desalt' a protein solution.

> **Key point**
>
> An analogy I have used in lectures: suppose a mixture of students—some drinkers and some non-drinkers—are walking down a street. Whenever the group passes a pub, the drinkers enter and have a beer; the non-drinkers just keep walking. The drinkers and non-drinkers will be separated.

Ion-exchange chromatography

Different proteins have different net charges. This depends on the amino-acid composition of surface-exposed residues. Is there a preponderance of negatively-charged sidechains, or of positively charged sidechains? Pack a column with beads to which charged molecules are attached. If the beads are negatively charged, the passage through the column of positively-charged proteins will be retarded. Conversely, if the beads are positively charged, passage of negatively-charged proteins will be retarded. Either choice allows separation of molecules with different charges.

Affinity chromatography

Suppose only one protein in a mixture binds to a particular ligand. Then a convenient way to purify the protein is to affix the ligand to a set of beads, pack a column with the beads, and run a mixture containing the target protein through the column. The target proteins will bind to the column—or at least be retarded in their passage through it. They are thus separated from other proteins that do not bind, and which will pass freely through the column. Subsequently, bound proteins must be released from the column, and collected. This allows fairly clean separation of proteins with the particular binding specificity.

Examples include:

1. The His-tag (also called 6xHis-tag) is one of the simplest and most widely used affinity purification tags, with six or more consecutive histidine residues attached to either chain terminus. These residues readily coordinate with transition metal ions such as Ni^{2+} or Co^{2+} immobilized on beads or on a resin, for purification.

2. Some proteins bind specific carbohydrates, useful as affinity tags.

3. It is also possible to use an antibody specific for the target protein as the ligand in an affinity column.

Specific expression of a target protein

With the determination of DNA sequences, it has become possible to insert a gene encoding a protein of interest into a cell, and cause the cell to express the protein copiously. This produces a starting mixture already enriched in the target protein, from which it is more easily isolatable. (A natural example of this is the overproduction of specific antibodies by myeloma patients. These served as sources of purified 'monoclonal' antibodies for research, before hybridoma and genetic engineering techniques became available.)

An advantage of producing a specific human protein in *E. coli* is that there will be no other *human* proteins as impurities. This is important in synthesizing human growth hormone for therapeutic purposes. Formerly, the source of human growth hormone was extraction from pituitary glands of deceased humans. The problem was that some of the extracted material was contaminated with prions, leading to Creutzfeldt-Jacob disease in patients. Using human growth hormone synthesized in *E. coli* eliminated this danger.

On the other hand, a problem with synthesizing human proteins in bacteria is that the bacterial cells cannot perform post-translational modifications, notably glycosylation. Use of mammalian cells to produce the desired proteins can overcome this deficiency.

Once you have isolated your protein, it would be very interesting to determine its structure.

3.2 Experimental methods of protein structure determination

X-ray crystallography

An atom excited by X-rays will reradiate the energy, uniformly in all directions. When X-rays excite atoms in a crystal, the reradiated waves from each individual atom interact with one another, producing a non-uniform pattern. Depending on the direction of observation, the reradiated waves may be in phase and reinforce one another (see Figure 3.5(a)); or be out of phase, in which case they will cancel to leave zero intensity (Figure 3.5(b)); or there may be any phase difference in between.

In a simple picture, if there are two emitters, the waves from each will reinforce if the path difference from the emitter to the point of observation is an integral number of wavelengths (see Figure 3.6).

In diffraction from a crystal, more than two waves are added up: each atom is the source of a component of the scattered wave. To achieve reinforcement of the waves reradiated from every unit cell of the crystal, the picture is that the atoms in the crystals lie on sets of parallel planes, which can be thought of as 'mirrors'. If the direction of observation of the reradiated beams correlates properly with the orientation of the planes occupied by atoms, the reemitted waves will all be in phase. This is Bragg's law: $n\lambda = 2d \sin\theta$ (see Figure 3.7). Here θ is the angle between the incident beam and the planes of atoms, d is the spacing between planes of atoms, and λ is the wavelength of the X-rays. The condition for reinforcement is that the difference in path length of waves reemitted by different planes of atoms must be an integral number (n) of wavelengths.

Experimentally, this means that the X-rays scattered from a crystal will form a pattern, reinforcing only in specific discrete directions. Figure 3.8

shows such a pattern, for sperm whale myoglobin (the first protein structure to be determined by X-ray crystallography).

The good news is that the intensity distribution in the X-ray diffraction pattern contains information sufficient to specify the individual atomic positions. The bad news is that it is difficult to extract this information. This is why the earliest protein crystal structure determinations were so challenging, and their success rightly regarded as *tours de force*.

The interpretation of the data produces a map of the electron density in the crystal. A crystallographer can fit a model into the **electron-density map**, from which the coordinates can be read off (see Figure 3.9).

 Key point

Which amino acid (from the canonical set of 20) has been rebuilt better into the electron density in Figure 3.9?

Fig. 3.5 (a) Waves that are 'in phase' reinforce each other (orange + blue = black). (b) Waves that are 'out of phase' can cancel each other out (orange + blue = 0). In frame (b) the blue curve is shifted relative to the orange curve by half a wavelength.

This figure shows the resultant intensity if the waves are either perfectly in phase or perfectly out of phase. If you were to do the corresponding calculations with two waves shifted by some random amount, you would in general find considerable remaining intensity, showing a more complex pattern. In **diffraction** from a crystal, each atom scatters X-rays. The larger the array of atoms, the more critical the dependence of scattering intensity on angle. For macroscopic crystals, containing very large numbers of atoms, the angular constraint for constructive interference is very precise, leading to a pattern of tiny discrete spots.

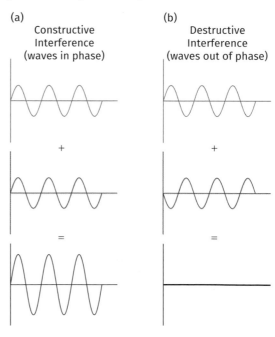

Fig. 3.6 Waves emitted from two sources (orange and blue) will reinforce each other if the difference in path length from the sources to the point of observation is exactly a wavelength (λ); or, more generally, an integral number of wavelengths.

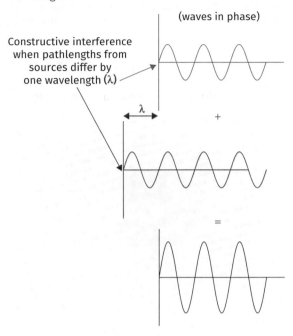

There are now mature methods for solving crystal structures of proteins, nucleic acids, and their complexes. Advances have involved both the instruments for recording the measurements, and the software used to interpret them. Many synchrotrons dedicate beams to protein crystallographic

Fig. 3.7 Bragg's law, $n\lambda = 2d\sin\theta$, expresses the condition for constructive interference, that the difference in path length between scattered X-rays from adjacent parallel planes must be equal to an integral number of wavelengths. d is the distance between the planes, and θ is the angle between the incident (and the reflected) rays and the scattering planes. The broken line has length d. The difference in distance travelled along the two paths, $2d\sin\theta$ is equal to the sum of the lengths of the orange segments, each of which is $d\sin\theta$.

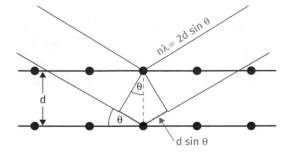

Fig. 3.8 Photograph of X-ray diffraction by a myoglobin crystal, from work by J.C. Kendrew and his collaborators. The distribution of the reflections in a diffraction pattern gives clues to the size and symmetry of the unit cell, and of the relative orientations of the subunits that it contains. The white splodge in the centre is the 'shadow' of the beam stop—a small bit of lead placed to interrupt the very high intensity of the unscattered beam.

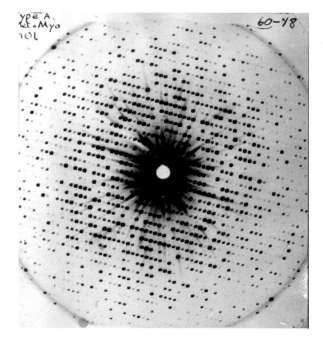

(Reproduced courtesy of the MRC Laboratory of Molecular Biology.)

Fig. 3.9 Fitting of a molecular model to an electron density. The network shows the experimentally-determined electron density, contoured at a suitable level. A crystallographer can manipulate the model to fit the electron density, using computer graphics. (a) Before rebuilding. (b) Rebuilding of the molecular model improves the fit to the electron-density map.

(a)   (b)

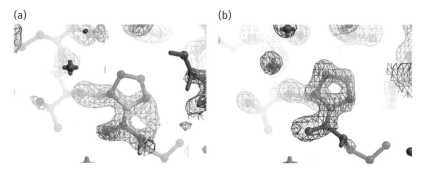

(From: Dodson, E. (2008). The befores and afters of molecular replacement. Acta Cryst., D64, 1724.)

data collection. These advances have supported an explosive growth in the contents of the databanks that archive and distribute macromolecular structures.

Indeed, at present solving structures is almost always quite straightforward, provided that it has been possible to produce crystals. But there is the problem. Some molecules prove difficult to crystallize. Membrane proteins are common examples, as they do not dissolve in typical aqueous media in common use in molecular biology. NMR spectroscopy and cryo-electron microscopy have the advantages of not requiring crystals.

Protein structure determination by nuclear magnetic resonance (NMR) spectroscopy

Most atomic nuclei have spins, which can adopt different energy levels in an external magnetic field. Nuclear magnetic resonance (NMR) spectroscopy measures the transitions between these energy levels. The energy differences of these transitions give clues to chemical structure.

- *The electrons surrounding the nuclei interact with the nuclear spins to modify the energy levels.*

The energy levels of a nucleus of non-zero spin depend on the local magnetic-field strength at the nucleus. This local field is a combination of the external applied field, and its perturbation by electrons in the vicinity of the nucleus. For example, the shielding of a proton in a methyl group will be different from that of a proton bound to a nitrogen.

NMR spectra therefore reflect the chemical environment of the nuclei, including in particular the bonds that the atoms form. The perturbation of the energy-level differences is called the **chemical shift**. Each type of chemical group will appear, shifted, in a different region of an NMR spectrum. In a protein, chemical shifts of nuclei of $C\alpha$ atoms reveal the secondary structure.

- *Excitation energy can transfer between different nuclei close together in space ($< \sim 5\text{Å}$ apart).*

From interactions through space between non-bonded atoms, NMR can identify pairs of atoms close together in the structure. Identification of residues distant in the sequence but close together in space is crucial to being able to assemble the polypeptide chain into the correct overall folding pattern.

The pattern of contacts allows solving for the structure. (The reader may think of a classic trestle railway bridge: The struts fix the distances between subsets of points, with the result that the structure is non-deformable (i.e. rigid); that is, the set of fixed distances determines the structure.) Figure 3.10 illustrates the nature of the measured information in protein structure determination.

Computer programs solve for structures compatible with the experimental data: the distance constraints, and the inferences about mainchain conformational angles from chemical shifts. The calculations include additional terms that build in proper stereochemistry, and ensure that the structures have low conformational energies. The most common technique for determining structure from NMR data is restrained molecular dynamics;

Fig. 3.10 A simplified picture indicating experimental information used by NMR spectroscopists to determine protein structures. This picture shows residues 20–40 and 72–103 from *Borrelia burgdorferi* outer surface lipoprotein. The entire protein is 122 residues long. Black lines indicate experimentally-determined pairs of residues close in space in the structure. (Other measured contacts between pairs of residues of these regions within five residues of each other in the sequence are not shown. Although this figure shows links between Cα atoms the experiment actually determines distances between protons.)

that is, molecular dynamics with additional terms that enforce agreement with the experimental data.

Whereas the result of an X-ray crystal-structure determination is usually a unique structure, NMR generally yields a set of similar but non-identical models (see Figure 3.11). There is some good-natured rivalry between specialists in different techniques—crystallographers scoff, saying that the NMR data simply underdetermine the structure. NMR spectroscopists retort that crystal-packing forces artificially constrain the dynamics of a genuinely-mobile structure.

Cryo-electron microscopy

Electron microscopy (EM) is the third experimental technique of protein structure determination. Recent breakthroughs have promoted it from a method that could produce only low-resolution structures—'blobs'—to the new successes that produce structures at atomic resolution (see Figure 3.12).

Shooting a beam of electrons through a molecule produces an image that integrates the electron density in the sample. Compare such an EM image with an ordinary photograph, such as a 'selfie' that you might take with your phone: the image from the phone shows only the *surface* of the subject. EM images are penetrative—as if your 'selfie' showed not only your face, but the bones of your skull, your brain, and the ponytail (if present) behind your head. It would still be a two-dimensional image, but contain full three-dimensional information.

How can we derive a true three-dimensional structure from such images? Suppose we disperse many copies of a molecule, or complex, on a grid on the stage of the microscope. We can observe many images of the same

Fig. 3.11 Models of *Drosophila* antennapedia protein determined by NMR spectroscopy. The core of the molecule, comprising the interacting regions of secondary structure, is consistent among the models, but the regions at the N- and C-termini are variable.

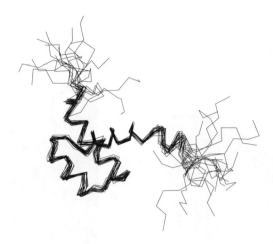

Fig. 3.12 Density map and atomic model for selected active site residues in *E. coli* β–galactosidase, determined by cryo-electron microscopy.

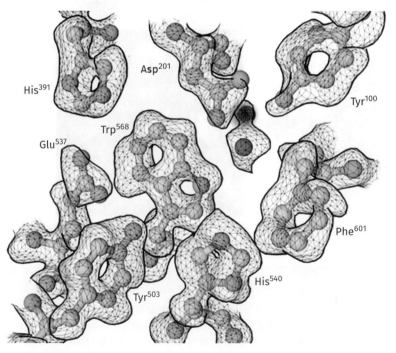

(Reprinted from Structure, 26/6, Alberto Bartesaghi, Cecilia Aguerrebere, Veronica Falconieri, Soojay Banerjee, Lesley A. Earl, Xing Zhu, Nikolaus Grigorieff, Jacqueline L.S. Milne, Guillermo Sapiro, Xiongwu Wu, Sriram Subramaniam, Atomic resolution cryo-EM structure of β-galactosidase, Pages 484–856.e3, Copyright (2018), with permission from Elsevier.)

object, but in different orientations. The challenge is to determine the orientations of the different images. But when this is achieved, we will be able to combine the images—different projections of the same object, in different orientations—into a three-dimensional structure.[1]

Figure 3.13 shows the overall course of a cryo-EM structure determination.

Fig. 3.13 Steps in a cryo-EM protein structure determination. An isolated protein is presented on a grid, producing an image containing many views of the structure. These are isolated (bottom row, right) and programs determine the orientation. The information from the combination of orientations is integrated into a structure.

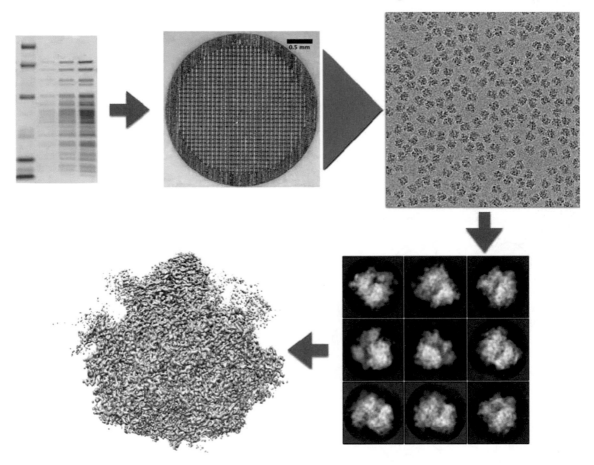

(Image courtesy of Chris Russo and Lori Passmore, MRC Laboratory of Molecular Biology.)

> 💡 **Key point**
>
> The three main experimental methods for determining protein structure are: X-ray crystallography, NMR spectroscopy, and cryo-electron microscopy. They are experimental in the sense that structure determination depends on physical measurements; however, all these methods are crucially dependent on the contributions of computer programs for data analysis.

---

[1]In the table game Foosball different copies of the same structures appear, on a lattice, in different orientations. See, for instance: https://www.123rf.com/photo_82959806_people-playing-enjoying-foosball-table-soccer-game-recreation-leisure.html

3.3   Protein structure prediction

We know that all the information necessary to specify the three-dimensional structures of proteins is contained in the amino-acid sequences. Nature has an algorithm that computes structure from sequence. If we could reproduce this algorithm in computer programs, we could predict protein structures. Indeed, *any* algorithm that works would solve the problem, even if it were not the algorithm that nature uses. The program AlphaFold2, from Google subsidiary DeepMind, has achieved spectacular success. There is consensus that this is a real 'game-changer'.

Critical Assessment of Structure Prediction—CASP

The Critical Assessment of Structure Prediction (CASP) programme organizes blind tests of protein structure predictions. Crystallographers, NMR spectroscopists, and cryo-electron microscopists make public the amino-acid sequences of the proteins they are investigating, and agree to keep the experimental structure confidential until predictors have had a chance to submit their models. After the deadline for submission of predictions, the experimental structures are released and the predictions evaluated.

CASP runs on a two-year cycle. Early in each active year, the target sequences are published, and predictions collected prior to release of the experimental structures. At the end of the year a gala meeting brings together the predictors and evaluators to discuss the current results and to assess progress. An album of papers appears as a supplement to the journal Proteins: Structure, Function and Bioinformatics.

Protein structure prediction is a classic challenge in computational biology. For many years—as recorded in successive CASP programmes—progress took place in fits and starts. Recently, however, very great progress has emerged.

Two approaches are sufficiently good to have practical application to academic and clinical research: homology modelling, and *a priori* prediction.

 Key point

Homology modelling is the prediction of an unknown structure, knowing the structures of one or more related proteins.

a priori structure prediction is the prediction of an unknown structure, without making explicit use of any other known structure. (However, many algorithms make use of *general* properties of known structures, such as the Sasisekharan-Ramakrishan-Ramachandran plot (Figure 2.7(b)), as well as amino-acid *sequences* of related proteins.)

Homology modelling

In homology modelling, one predicts the structure of a protein from the known structures of related proteins. The task is to predict the *changes* in structure between the known structures used in the modelling, and the actual structure of the target. One might consider *a priori* prediction 'from scratch' as the *integral* form of the protein structure prediction problem, and homology modelling as the *differential* form.

Figure 3.14 shows the **superposition** of hen egg white lysozyme, and baboon α-lactalbumin. Suppose you knew the amino-acid sequence and three-dimensional structure of hen egg white lysozyme, but only the amino-acid sequence of baboon α-lactalbumin. Proposing the chain tracing of lyzosyme, as shown in the Figure, as a model for the structure of α-lactalbumin would give a pretty good answer.

To go beyond this, you would have to improve the prediction of the regions in which there were structural differences between hen egg white lysozyme and baboon α-lactalbumin. (Of course you would not know where in the structure these differences appear, as we are assuming that you do not know the α-lactalbumin structure.)

Modern homology modelling methods are quite sophisticated and successful. They make use not only of a single similar structure, but of a whole family of similar structures. The challenge is to identify the most useful features of each source structure, and to combine them into the best possible prediction. The program **I-TASSER**, by Y. Zhang and colleagues of the University of Michigan, U.S.A.,[2] has consistently provided superb results (see Figure 3.15).

a priori structure prediction

Recently, a method of structure prediction *not* based on the structures of specific known relatives of the target protein has achieved a major and spectacular breakthrough.

In the latest completed CASP programme (CASP14, 2020), a program **AlphaFold2** by Google subsidiary DeepMind showed phenomenal success. The approach made use of techniques of artificial intelligence, a field made famous by computer programs able to beat the human world champions in Chess and Go.

Fig. 3.14 A superposition of hen egg white lysozyme (blue) and baboon α-lactalbumin (magenta). There are only 37% identical residues in an alignment of the amino-acid sequences, but the structures are nevertheless quite similar.

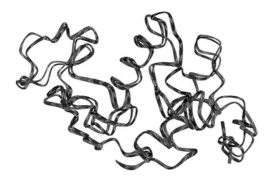

[2]Zheng, W., Zhang, C., Bell, E.W. and Zhang, Y. (2019). I-TASSER gateway: A protein structure and function prediction server powered by XSEDE. Future Gener. Comput. Syst. 99, 73–85.

Fig. 3.15 Homology-modelling by I-TASSER of target T1006 from CASP 13.

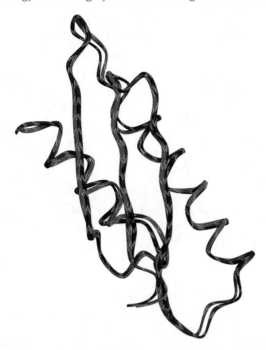

The magazine Nature, usually characterized by a relatively restrained tone of expression, published on 10 December 2020 a commentary article with the banner—and fully upper case!—headline:

'IT WILL CHANGE EVERYTHING': AI MAKES GIGANTIC LEAP IN SOLVING PROTEIN STRUCTURES

DeepMind's program for determining the 3D shapes of proteins stands to transform biology, say scientists.

Figure 3.16 shows the predictions by AlphaFold2 of two targets from CASP14:

- T1038: Domain 2 of the tomato spotted wilt tospovirus glycoprotein
- T1049: Domain 1 of the major virulence-associated fimbrial protein, MrpH, of the bacterial urinary tract pathogen *Proteus mirabilis*.

The results are spectacular: The chain tracings of the experimental result and the prediction are virtually identical. To draw the pictures, it was necessary to reduce the diameter of the chain shown in the representation, in order to see any differences at all.

What is more, these examples are not selected rare successes; AlphaFold2 achieved comparable results consistently. Nearly two-thirds of its predictions are comparable in quality to experimentally-determined structures.

Fig. 3.16 (a) Prediction by AlphaFold2 of domain 2 of target T1038 from CASP 14, tomato spotted wilt tospovirus glycoprotein. (b) Prediction by AlphaFold2 of domain 1 of target T1049 from CASP 14, major virulence-associated fimbrial protein, MrpH, of the bacterial urinary tract pathogen *Proteus mirabilis.*

(a)   (b)

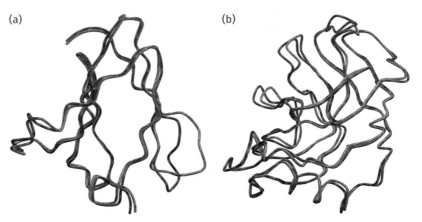

Indeed, John Moult, the originator and leader of the CASP programs since its inception in 1994—and who must deservedly be dead chuffed at the triumph of the effort—said, 'In some cases, it was not clear whether the discrepancy between AlphaFold's predictions and the experimental results was a prediction error or an artefact of the experiment.'

For example, in Figure 3.16(a), the conformation of a long loop at the lower right differs between prediction and experiment, much more than any other region of the structure. In the experiment, the corresponding residues are involved in interactions between different molecules in the crystal. Such crystal-packing contacts are known, in many cases, to perturb the conformations of the regions involved, in the resulting experimental structure. It may well be that, in this case, the predicted structure is a more accurate statement of the solution conformation than the X-ray crystal structure. (How could you decide whether this is true?) One clue is available from the AlphaFold2 calculation itself: In addition to the structure prediction, AlphaFold reports the expected reliability of different parts of each predicted protein structure, based on an internal confidence measure. For this prediction, the error estimates for the residues in this loop are above average, but do not stand out as much larger than other regions of the structure.

The AlphaFold program improved between CASP 13 and CASP 14. The team stated in November 2020 that 'they believe AlphaFold can be further developed, with room for further improvements . . . .'

The program takes as input data a query amino-acid sequence, used initially to gather a set of similar sequences from databanks, to form a multiple-sequence alignment. Patterns of residue conservation shown in such alignments contain many clues to the three-dimensional structure. For instance, positions at which residue variability is correlated suggest residues that are nearby in three-dimensional space. This is the type of information that is measured by NMR spectroscopy (see Figure 3.10).

How does the method work? AlphaFold2 deals with (1) the distance matrix, that is the set of interresidue distances, including but not limited to indicating pairs of residues that are close together in space, and (2) the conformational energy of different conformations, based on force fields that include terms representing bond lengths and angles, torsional angles, and interactions among non-bonded interactions.

AlphaFold2 processes this information by a machine-learning technique called a 'neural network'. There are multiple, intersecting pathways of information transfer from the input data to the prediction. In a rough simulation of the real neural networks in animal brains, the network is reconfigurable, and the strengths associated with different pathways through it can be altered. These changes in the network, analogous to 'learning', will alter the output—the conclusion from the data, in the form of a predicted structure. 'Training' allows calibration and refinement of the network. The mature network is then applicable to prediction of protein structure from amino-acid sequence.

Interestingly, the use of multiple-sequence alignments means that the method has not yet been proven to achieve the 'Give me ONLY a single amino-acid sequence and I'll tell you the structure' power that real proteins achieve. For instance, one might wonder about predicting the structure of a designed protein that has no natural homologues. Would AlphaFold2 (not to mention other methods) perform more poorly in such cases?

Nevertheless, there is consensus that, as far as pure science is concerned, this is one of the major achievements of 2020. A spate of papers has recently appeared in Nature and in the Journal of Molecular Biology. Indeed, news has spread beyond the specialist literature; see, for example:

https://www.theguardian.com/technology/2020/nov/30/deepmind-ai-cracks-50-year-old-problem-of-biology-research

https://www.economist.com/science-and-technology/2020/11/30/how-do-proteins-fold

https://www.youtube.com/watch?v=5YRVhJHDcyM

and even

https://www.youtube.com/watch?v=ZHfumZVPjVA

Key point

The program AlphaFold2, from Google subsidiary DeepMind, has recently shown spectacular success. There is consensus that this is a real 'game-changer'.

 ## Summary Points

- Study of a protein begins with isolating it from a natural source containing the target protein as a component of a mixture. Methods for protein purification include: Ammonium sulphate precipitation; size-exclusion chromatography; ion-exchange chromatography; affinity chromatography; and specific expression, often in *E. coli*.

- The major experimental methods for determining protein structures are X-ray crystallography, NMR spectroscopy, and cryo-electron microscopy.
- Computational methods for protein structure prediction from amino-acid sequence, have recently shown very great improvements. A truly general and reliable method of structure prediction will revolutionize research on proteins.

Further Reading

Bonner, P.L.R. (2019). Protein Isolation. 2nd edn (Boca Raton, FL: CRC Press).
A good general source.

Grimes, J.M., Hall, D.R., Ashton, A.W., Evans, G., Owen, R.L., Wagner, A., McAuley, K.E., von Delft, F., Orville, A.M., Sorensen, T., Walsh, M.A., Ginn, H.M., Stuart, D.I. (2018). Where is crystallography going? Acta Crystallogr. D. Struct Biol. 74(Pt 2): 152–66.
New developments in X-ray crystallography.

Wüthrich, K. (1995). NMR–this other method for protein and nucleic acid structure determination Acta Cryst. D51, 249–70.
Still very useful.

Callaway, E. (2020). The protein-imaging technique taking over structural biology. Nature 573, 201.
Comment on state-of-the-art in cryo-electron microscopy, and its implications.

Cheng, Y. (2018). Single-particle cryo-EM—How did it get here and where will it go. Science. 361, 876–80.

Nwanochie, E. and Uversky, V.N. (2019). Structure determination by single-particle cryo-electron microscopy: only the sky (and intrinsic disorder) is the limit. Int. J. Mol. Sci. 20, 4186.
Two papers on recent developments in cryo-electron microscopy.

Greener, J.G., Kandathil, S.M. and Jones, D.T. (2019). Deep learning extends de novo protein modelling coverage of genomes using iteratively predicted structural constraints. Nat. Commun. 10, 3977.
A recent review of new methods of structure prediction.

Kryshtafovych, A., Schwede, T., Topf, M., Fidelis, K. and Moult, J. (2021). Critical Assessment of Methods of Protein Structure Prediction (CASP)-Round XIII. Proteins: Structure, Function, Bioinformatics in press. https://predictioncenter.org/casp14/doc/CASP14_Abstracts.pdf
Report on latest CASP programme.

Kuhlman, B., Bradley, P. (2019). Advances in protein structure prediction and design. Nat. Rev. Mol. Cell. Biol. 20, 681–97.
A fairly recent publication, but the field is moving very very fast.

Renaud, J.-P., ed. (2020). Structural Biology in Drug Discovery: Methods, Techniques, and Practices (Hoboken, NJ: J. Wiley & Sons).
A comprehensive book, some of which covers material in this chapter.

Bricogne, G., 'High-Throughput Macromolecular Crystallography in Drug Discovery: Evolving in the Midst of Revolutions'. (*Anything* written by Bricogne will repay study.)
and

Delsuc, M.-A., Vitorini, M. and Kieffer, B. 'Determination of Protein Structure and Dynamics by NMR: State of the Art and Application to the Characterization of Biotherapeutics'.
Two chapters in the book edited by Renaud (above) are particularly relevant.

Deane, C. (2020).
https://www.blopig.com/blog/2020/12/casp14-what-google-deepminds-alphafold-2-really-achieved-and-what-it-means-for-protein-folding-bi-ology-and-bioinformatics/
Several papers have appeared in July 2021 issues of Nature:

Callaway, E. (2021). DeepMind's AI predicts structures for a vast trove of proteins. doi: 10.1038/d41586-021-02025-4

Jumper et al. (2021). Highly accurate protein structure prediction with AlphaFold. doi: 10.1038/s41586-021-03819-2

Tunyasuvunakool, K. et al. (2021). Highly accurate protein structure prediction for the human proteome. doi: 10.1038/s41586-021-03828-1

Descriptions of AlphaFold, a program for a priori *protein structure prediction.*

Discussion Questions

3.1. For many years, it was believed that amino-acid sequences folded into native states in which every residue assumed the same conformation in all copies of the protein. (That haemoglobin adopted *two* rather than one state was considered an exception that proved the rule.) Now we understand that parts, or even all, of some proteins, can be *intrinsically disordered*. Compare the methods of structure determination of proteins in terms of what they can reveal about intrinsic disorder in protein structures.

3.2. How will the development of general and reliable methods of structure prediction affect investigations of proteins? Consider the current disparity between the extents of our knowledge of amino-acid sequences and of protein structures.

3.3. Suppose you wish to design an artificial protein with a novel enzymatic activity. Which principles and methods discussed in this chapter would you use in addressing this problem?

4 PROTEIN FUNCTION

Learning Objectives

- Appreciate the very wide variety of ways that proteins support the structures and life processes of viruses, cells, and organisms.

- Know types of structural proteins and their roles in biological architecture.

- Understand the basic principles of enzymatic catalysis by proteins, as presented in the distribution of thermodynamic changes in a reaction diagram. Be able to interpret K_M and V_{max}, the values that characterize enzymatic activity. Know the nature of the experiments used to determine them.

- Know the basic structure of immunoglobulins, and understand how the diversity of their binding sites allows them to recognize the organic world. Recognize their utility in clinical applications for diagnosis and therapy.

- Appreciate that phospholipid bilayers form the basic structure of the membranes that surround cells and intracellular organelles, and recognize that control of traffic across membranes requires membrane-embedded proteins that function as channels or pumps. Understand how the arrival of a signal at a cell surface can inform the cell interior, to initiate a control cascade, without the signalling molecule itself needing to enter the cell.

- Be able to summarize the types and mechanisms of free-energy changes involved in the basic steps of digestion of glucose—glycolysis, the Krebs cycle, the electron-transport chain, and ATP synthase—leading to the ultimate capture of energy in the form of ATP.

- Know the design of the protein-classification schemes of the Enzyme Commission and the Gene Ontology projects.

Protein structure governs protein function. The great variety of protein functions demands great variety of protein structures. The challenges to us are: (1) to derive from structure the detailed mechanism of action. (2) to *predict* the functions of proteins. This is necessary for annotation of newly-sequenced genomes. (3) to design novel sequences which will form proteins with novel functions.

It is useful to have catalogues of protein functions. Then, to assign function of a protein, it is not necessary to compose a description of the activity, but only to place it in the right 'pigeonhole'. The **Enzyme Commission** and **Gene Ontology** projects provide such classification schemes.

Examples of structure-function relationships discussed in this chapter include:

- Structural proteins
- Enzymes—catalytic proteins
- Antibodies
- Transmembrane transport proteins, including channels, gates, and pumps; and receptors, all embedded in the general **phospholipid bilayer** membrane structure
- The role of transmembrane proton gradients in the electron transport chain and ATP synthesis

4.1   Structural proteins

Structural proteins have the role of solids in the world of proteins. Some provide rigidity: either for purely architectural purposes; or for mechanical applications, serving as fulcra. A common feature is that they are not soluble in water, for obvious reasons.

Some structural proteins are fibrous. When you look at other people, what you see is mostly **fibrous protein**. Hair contains α–keratin, a rope assembled from coiled-coils of α–helices (see Figure 4.1). The helical section of a single subunit of α–keratin typically contains ~300 residues in each of the two strands, and is 48nm long. There are also small capping domains at head and tail. These units assemble into large aggregates, extensively cross-linked by disulphide bridges. The 'permanent wave' treatment of hair first breaks these disulphide cross-links using a reducing agent. Then, after setting the hair into

Fig. 4.1 Coiled-coiled helices from α–keratin, the basis of the structure of human and animal hair and other structural proteins. Each helix is structurally similar to a standard α–helix, but whereas a standard α–helix is straight, these helices are 'supercoiled', winding around each other. Hair contains many such molecules, packed side-to-side to form the macroscopic fibres. The molecules interact by hydrogen bonding and by disulphide bridges.

Claws, beaks, and feathers of birds contain a different molecule, called β–keratin.

the desired macroscopic conformation, the links are reformed by a mild oxidizing agent, usually dilute hydrogen peroxide, preserving the chosen 'hair-do'.

The horny outer layer of the skin, and fingernails, are other forms of keratin, differing in amino acid composition and sequence. Fingernails are less flexible than skin or hair because of a greater abundance of cysteine residues forming disulphide cross-links.

The layer of the skin just beneath the surface contains collagen, a glycoprotein (see Case study 4.1). The cornea of the eye is another form of collagen. Fibrous proteins are ubiquitous beneath the skin also; for instance, tendons contain collagen. Indeed, collagens make up about a third of the protein content of the human body.

Many traditional clothing materials—before the development of synthetic fibres—are also protein. Woollen clothes are α–keratin—made from the hair of sheep or other animals. (Cotton and linen, in contrast, are plant fibres, formed primarily of cellulose.) Animal hair is of course chemically similar to human hair, but varies in physical properties among species. (Paintbrushes have been made from the fur of many animals, including sable, ox, squirrel, pony, goat, hog, camel, and mongoose. Artists are exquisitely sensitive to the variations in stiffness, and the differences in retention and delivery of paint.)

Silk is β–fibroin, with the repetitive sequence . . . Gly-(Ala or Ser)-Gly-(Ala or Ser) . . ., forming an extended β–sheet. The cocoons of moths contain fibres of β–fibroin glued together by a second protein, sericin. When ready to emerge, moths secrete the proteolytic enzyme cocoonase to dissolve the sericin and let them out. Cocoonase can also digest proteins other than sericin, including keratin. It is a paradox that moths can digest the keratin of woollen sweaters, and the fabric of their own cocoons, but not silk scarves! The explanation is that most silk cloth contains the β–fibroin but not the sericin.

 Key point

Fibrous proteins contain assemblies of α–keratin, collagen, and fibroin. Control over degrees of cross-linking determines hardness; compare fingernails and hair, both formed of α–keratin.

Case study 4.1
Collagen

The many different types of collagen have a common basic structure, a triple-stranded helix, with three polypeptide chains coiled around one another in a plait (see Figure 4.2). Individual chains are each ~1,000 amino acids long. They have a glycine every third residue—the sequence scheme is $(Gly–X–Y)_n$. X and Y are mostly alanine or special modified amino acids: X is often a 3- or 4-hydroxyproline and Y a hydroxyproline or hydroxylysine.

Fig. 4.2 The structure of collagen, a three-stranded supercoil formed by plaiting together three polypeptide chains. Each polypeptide chain itself forms a helix, with approximately 3.3 residues per turn. (This helix is different, structurally, from an α–helix.)

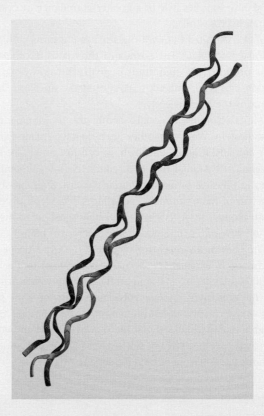

Collegen molecules assemble into fibrils in different ways, suitable for the differing mechanical requirements of the tissues in which they appear. Forty-two types of human collagen have been distinguished, encoded by genes on several different chromosomes. The differences in amino acid composition and sequence govern their modes of processing and assembly. Hydroxylysines provide sites of linkage to sugars or short polysaccharides; differing amounts of hydroxylysine account for the different carbohydrate content of different collagen types. Alternatively, covalent cross-links form between strands, after lysine or hydroxylysine sidechains have been oxidized enzymatically to aldehydes. These aldehydes react either with other aldehydes or amino groups of lysine or hydroxylysines on neighbouring collagen chains. The cross-links contribute to the mechanical strength of the fibre.

Several genetic abnormalities affect the structure of connective tissue. In one form of **Ehlers-Danlos syndrome**, mutations in the gene for the enzyme lysine hydroxylase cause defective post-translational modification of lysine to hydroxylysine. The consequent reduction in cross-linking lowers the mechanical strength. The symptoms of Ehlers-Danlos syndrome include spidery fingers and unusually high flexibility of the joints. The nineteenth-century violinist Niccolò Paganini exhibited these anatomical characters—and to very good use he put them! It has been speculated that he had either Ehlers-Danlos syndrome or **Marfan syndrome**, a condition affecting another connective tissue protein, fibrillin.

Except for most primates, many animals can synthesize vitamin C and do not need to obtain it from their diets. Humans lack a single enzymatic activity in the synthetic pathway from glucose to vitamin C: L-gulonolactone oxidase, the enzyme that catalyses the last step in the pathway.

The symptoms of **scurvy**, the deficiency disease caused by lack of vitamin C, result from defective hydroxylation of prolines and lysines, weakening collagen. Vitamin C is an essential cofactor of prolyl and lysyl hydroxylases. Loss of integrity of gum tissue and of dentin, inside teeth (dentin forms by mineral deposition on a collagen matrix), leads to loosening and ultimately falling out of teeth.

Sailors on diets restricted by what could be carried on long voyages traditionally suffered from scurvy. An eighteenth-century Scottish surgeon, James Lind, discovered that citrus fruit could prevent the disease. This is the origin of the term 'limey', referring originally to British sailors, and, by extension, to the entire nation.

Discussion Questions

1. There are five major types of collagen in the human body. What are their structural differences? In which parts of the body are they major structural components? How do their structural differences support their different functions?
2. Collagen fibres from the intestines of animals have been used as strings, both for violins and other 'stringed' instruments; and for tennis rackets. They are colloquially called 'gut strings'. Originally there was no alternative, but now metal and synthetic polymer strings are available, and in much more common use. What are the arguments for and against using traditional collagen strings? What properties of the collagen fibres are involved?

4.2   Enzymes

Enzymes are treated thoroughly in another volume in this series. Therefore the following discussion is limited to essential principles.

Proteins that catalyse reactions are enzymes. Unlike general non-biological catalysts, most enzymes show specificity of two kinds:

1. Enzymes catalyse the reactions of only a limited range of substrates, often only a single substrate.

Key point

Enzymes do not affect equilibrium constants, but speed up the approach to equilibrium.

2. Enzymes catalyse only a specific reaction of the substrates on which they act. For instance, pyruvate is a branch compound in metabolic pathway networks; that is, different biochemical reactions of pyruvate can produce approximately 30 different products. A different enzyme is needed to catalyse each of these reactions.

In addition to proteins, some enzymes are RNA molecules; these are called ribozymes. Even some DNA sequences show catalytic activity.

 Key point

Emil Fischer explained enzyme specificity in terms of a 'lock-and-key' model, hypothesizing structural complementarity between enzymes and substrates. This explains substrate specificity, but not reaction specificity.

How do enzymes speed up reactions?

A typical enyzme-catalysed reaction proceeds through the following steps:

Enzyme + Free Substrate E + S →
  Enzyme-Substrate complex ES →
    Enzyme-Substrate transition state ES‡ →
      Enzyme-Product complex EP →
        Enzyme + Free Product E + P

or:

$$E + S \rightarrow ES \rightarrow ES^{\ddagger} \rightarrow EP \rightarrow E + P$$

Figure 4.3 shows the variation in standard Gibbs Free Energy ($G^{\ominus}$) of the different stages of the process.

 Key point

The standard Gibbs Free Energy ($G^{\ominus}$) is the relevant thermodynamic quantity to describe an enzymatic reaction. Differences in $G^{\ominus}$ determine equilibrium constants, and reaction rates.

- Green arrow = energy of stabilization of Enzyme-Substrate complex (related to the Michaelis constant, K_M).
- Light-blue arrow = activation energy from Enzyme-Substrate complex to Transition State. This quantity determines the rate of the forward reaction.

Fig. 4.3 Standard Gibbs Free Energy ($G^{\ominus}$) relationships in different stages of a typical enzyme-catalysed reaction.

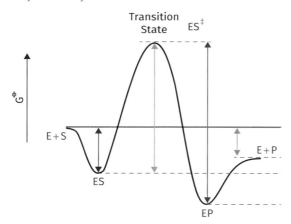

- Dark-blue arrow = activation energy from Enzyme-Product complex to Transition State. This quantity determines the rate of the reverse reaction.

- Purple arrow = Gibbs Free Energy change between Substrate and Product (the Gibbs Free Energy for free Enzyme E is the same in both E + S and E + P states). This overall Gibbs Free energy change determines the equilibrium constant K_{eq} for the overall reaction S → P; according to the equation: $\Delta G^{\ominus} = -RT$ in K_{eq}, in which: $\Delta G^{\ominus}$ = change in Gibbs free energy, under standard conditions, R = gas constant = 8.314 J·K⁻¹· mol⁻¹ (in which K = absolute temperature). (Do not confuse K = absolute temperature with K_{eq} = equilibrium constant!)

Key point

1. The enzyme cannot change the equilibrium constant of the overall re-action; therefore the difference in $G^{\ominus}$ levels between the start and end states S and P is independent of the enzyme.

2. A stronger affinity in the Enzyme-Substrate complex (longer green arrow) will lower the level of ES, and thereby *raise* the activation energy (giving a longer light-blue arrow). If the enzyme is to speed up the reaction, relative to the uncatalysed rate, the affinity of the enzyme for substrate must be *less than* the affinity of the enzyme for the transition state.

Enzyme kinetics

L. Michaelis and M. Menten developed a model for the kinetics of enzyme-catalysed reactions. The experiment envisaged was to mix a fixed amount of enzyme with different amounts of substrate (in separate tubes) and to measure the initial velocity v_0 of the reaction.

The enzyme-substrate complex can form, but the reaction will not take place. In the Michaelis-Menten model, K_M is the equilibrium constant for the dissociation of the enzyme-substrate complex. Therefore a non-competitive inhibitor will not change K_M, but will of course reduce V_{max}. In contrast, a competitive inhibitor *will* change K_M, but not change V_{max} (because, at sufficiently high substrate/inhibitor concentration ratios, the effect of the inhibitor will be 'washed out'). These differences allow determination of the nature of an inhibitor—competitive v. non-competitive—from the enzyme kinetics.

Allosteric changes are a more complex method of control through ligand binding. Like non-competitive inhibitors, allosteric ligands bind away from active sites. However, they can either reduce activity—as with non-competitive inhibitors—or increase it. To influence events at the active site, allosteric control involves conformational change.

2. *Control over expression of the gene coding for the enzyme.* This affects the total amount of enzyme present, and adjusts the throughput rate accordingly.

3. *Post-translational modification.* For instance, the intracellular domains of many signal-receptor proteins are activated by tyrosine phosphorylation.

 Key point

Cells must control and organize the activities of their enzymes. Mechanisms include inhibition and allosteric change, control over expression, and post-translational modification. (See Fig. 1.6.)

4.3   Antibodies

Antibodies are a large family of proteins, tasked with protecting vertebrates against foreign molecular toxins, and infections by pathogenic bacteria and viruses. Within our bodies, antibodies form a very large family that are able, as an ensemble, to recognize the entire organic world. Genetic mechanisms create diversity in the immune system, such that our bodies contain something like 10^{10} antibodies. Moreover, in response to a novel antigen—for instance a new virus—we can produce novel antibodies to neutralize the new threat.

 Key point

There is a race between our immune systems and the effects, potentially fatal, of an infection. Sometimes the virus wins.

The discovery of methods for preparing single specific antibodies from this swarm—called a **monoclonal antibody**—has made possible many significant applications in clinical practice to diagnosis and therapy.

A typical antibody of the IgG type contains four polypeptide chains, two identical **light chains** and two identical **heavy chains**. The proteins are dimers of dimers. Both light and heavy chains contain copies of homologous domains, about 110 residues in length. The structure of each domain contains two antiparallel β–sheets, packed face-to-face. Each light chain comprises two domains; each heavy chain comprises four domains. Figure 4.5 shows their assembly into a Y-shaped structure. Two identical antigen-binding sites appear at the 'wing-tips' (* in Figure 4.5).

Fig. 4.5 (a) The structure of a typical antibody, type IgG. It is a tetramer (a dimer of dimers) containing two relatively short (low molecular weight) *light chains* (red and green), and two longer (higher molecular weight) *heavy chains* (cyan and magenta). Each light chain contains two domains; each heavy chain contains four domains. There are two identical antigen-binding sites, located at the 'wingtips' (marked by *). (b) Schematic diagram of the domain structure and chain assembly of an IgG. Light chains contain two domains: a variable domain V_L and a (relatively) constant domain C_L. Heavy chains contain four domains, a variable domain V_H, and three (relatively) constant domains C_H1, C_H2, and C_H3.

(a)

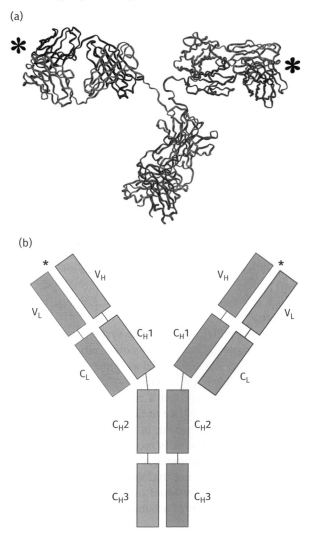

(b)

the rat antibody replace the original human CDRs. This produces a molecule which is *almost* completely human—and hence has reduced antigenicity in patients—but retains the binding affinity and specificity of the rat antibody. Such **humanized antibodies** have proved successful in treating patients. Perhaps the best known is **herceptin**, an antibody that blocks growth-factor receptors on the surfaces of breast-cancer cells. (Some antibodies useful for therapy are now produced from genetically-modified mice in which the antibody-coding regions of the mouse genome are replaced by human equivalents. Such mice produce *human* antibodies.)

> ### 💡 Key point
>
> Humanization is an example of antibody engineering, which also includes development of catalytic antibodies. This is logical: for an enzyme, it is necessary to have a molecule that binds substrate and juxtaposes bound substrate to catalytic residues from the enzyme. Antibodies solve part of this problem by providing binding sites.

4.4 Membrane proteins and receptors

Membranes are the frontiers of cells, and of intracellular organelles such as chloroplasts and mitochondria. More than passive wrappers, membranes control the flow of substances and information across the boundaries.

Membrane structures are based on phospholipid bilayers. It is essential that (1) the membrane *not* be water soluble, but (2) it must interact with aqueous phases at both the inner and outer surfaces. A typical component of a membrane, phosphatidylethanolamine (see Figure 4.8) has both charged and hydrophobic parts.

An assembly of phosphatidylethanolamine and related molecules forms a two-dimensional double layer 'sandwich'. In both layers, the hydrophobic tails point into the membrane and the charged heads point out (see Figure 4.9).

Fig. 4.8 The structure of phosphatidylethanolamine, a component of membranes. Phosphatidylethanolamine is a derivative of glycerol ($HOCH_2 - CHOH - CH_2OH$ = 1,2,3-trihydroxypropane), in which two of the three hydroxyl groups are esterified with fatty acids, to give a hydrophobic 'tail' of the molecule; and the third hydroxyl is esterified with phosphoric acid, derivatized with ethanolamine, to give a charged 'head' of molecule. In this figure the glycerol moiety appears in red.

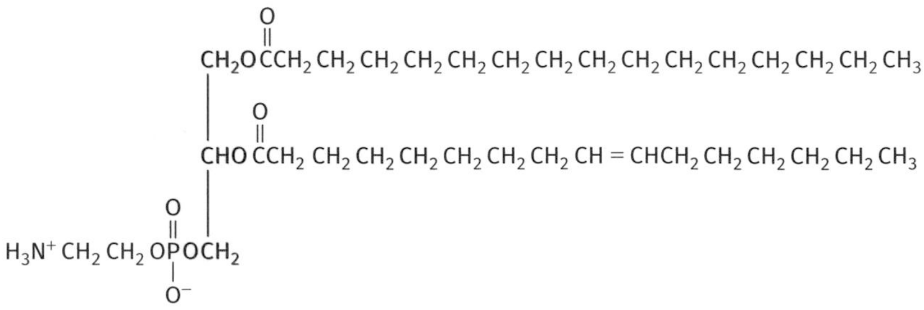

$$\begin{array}{l} \quad\quad\quad\quad\quad\ \ \overset{O}{\overset{\|}{}} \\ CH_2O\,CCH_2\,CH_2\,CH_2\,CH_2\,CH_2\,CH_2\,CH_2\,CH_2\,CH_2\,CH_2\,CH_2\,CH_2\,CH_2\,CH_3 \\ |\quad\quad\quad\ \ \overset{O}{\overset{\|}{}} \\ CHO\,CCH_2\,CH_2\,CH_2\,CH_2\,CH_2\,CH_2\,CH_2\,CH=CHCH_2\,CH_2\,CH_2\,CH_2\,CH_2\,CH_3 \\ \quad\quad\quad\ |\ \\ \overset{O}{\overset{\|}{}}\ | \\ H_3N^+\,CH_2\,CH_2\,O\,POCH_2 \\ \quad\quad\quad\quad |\ \\ \quad\quad\quad\quad O^- \end{array}$$

Fig. 4.9 Schematic diagram of the structure of a membrane. The phospholipids form two two-dimensional arrays: the 'lipid bilayer'. Each phospholipid is simplified to show a sphere for the charged head, pointing out of the membrane on either side; and the two fatty acid components of the tail, pointing into the membrane. Many other molecules are embedded in the membrane. These include pores and signal-receptor molecules which often have a structure containing seven trans-membrane helices.

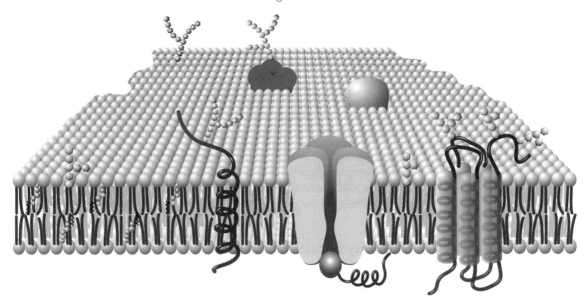

 Key point

The phospholipid bilayer is impermeable to many water-soluble compounds that must enter and exit the cell or organelle. Proteins embedded in the membrane provide routes of access. They must also exert control: (1) They must have specificity over what molecules they allow to pass through, and (2) They must control the direction of flow. In many cases channels have 'open' and 'closed' states that allow or forbid passage.

4.5   Transport proteins

Cells have the general problem of getting molecules from here to there. **Transport proteins** facilitate the delivery of molecules around a cell or an organism. The origin and destination can be local, as in transport across a membrane. They can be within a cell, as in the transport of organelles along cytoskeletal tracks, by **kinesins** and **dyneins**—these proteins bind cargo and carry it along the microtubules to a destination. Other proteins can carry ligands around the entire body, in the bloodstream. Examples include proteins that transport cholesterol (necessary because the cargo is not water-soluble) or retinol, and—the classic example—transport of oxygen and carbon dioxide by haemoglobin.

Fig. 4.11 Electron micrograph of mitochondrion from bat pancreas. Notice the very narrow width of the intermembrane space.

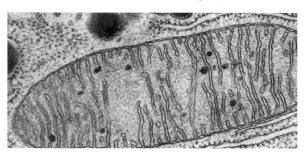

(K.R. Porter/Science Photo Library.)

 Key point

Recall that oxidation-reduction reactions involve electron transfers.

As electrons pass along the components of the electron-transport chain, cofactors of these proteins are transiently reduced (as they accept the electrons), and then reoxidized (as they pass the electron on to the next molecule in the chain) (see Figure 4.12). These spontaneous oxidation-reduction reactions are coupled to the pumping of protons, H^+, from the inner mitochondrial matrix, into the space between the inner and outer mitochondrial membranes. This creates a concentration disequilibrum across the inner membrane, with higher $[H^+]$ in the intermembrane space. (The pH difference can be about 1 pH unit; that is, a 10-fold concentration disparity.) Note that the transfer of protons from a region of low concentration to a region of high concentration is a non-spontaneous process. It must be driven by the free energy released in the oxidation-reduction reactions of the electron-transport chain.

By this mechanism the free energy of spontaneous electron transport is captured in the proton concentration disequilibrium. The next step is to couple the spontaneous passage of protons back into the mitochondrial matrix, to ATP synthesis. This is achieved by a small molecular motor, ATP synthase.

 Key point

ATP synthase couples the reentry of protons into the inner mitochondrial matrix to the synthesis of ATP. The proton transfer, *with* the concentration gradient, is a spontaneous process, which drives the non-spontaneous phosphorylation of ADP: $ADP + P_i + ATP$.

ATP synthase

ATP synthase contains a microscopic rotatory motor (see Figure 4.13). It acts as a 'turnstile', in that passage of protons through it drives the rotation. In its overall structure, ATP synthase is shaped like a mushroom (see Figure 4.14).

Fig. 4.12 Proton translocation in the mitochondrial electron-transport chain. Glycolysis (in the cytoplasm, not shown) and the Krebs cycle (within the mitochondrion) produce NADH and $FADH_2$. The reoxidation of these cofactors is coupled to passage of electrons down an electron transport chain. The ultimate electron acceptor is oxygen, reduced to water. The free energy of these spontaneous electron transfers is captured by producing a proton concentration disequilibrium between the inner mitochondrial matrix and the space between the inner and outer membranes. To reenter the mitochondrial matrix, spontaneously relieving the concentration disequilibrium, the protons pass through ATP synthase, converting the free energy of the concentration disequilibrium to formation of ATP. (This schematic diagram does not accurately represent the very large volume difference between the mitochondrial matrix and the intermembrane space.) https://commons.wikimedia.org/wiki/File:Mitochondrial_electron_transport_chain_(annotated_diagram).svg

Mitochondrial electron transport chain

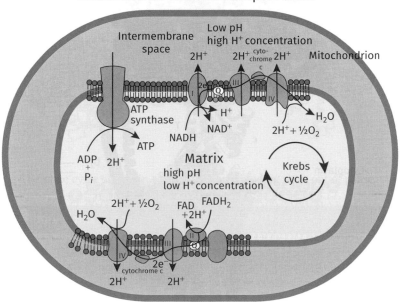

The bottom of the 'stem', containing the axle of the motor, is inserted into the membrane. The 'cap' of the mushroom contains the active sites for catalysis. It is a trimer of dimers, with binding sites between the two molecules of each dimer. As the axle rotates, it forces conformational changes in the active sites to drive the reactions.

Paul D. Boyer originally proposed this mechanism. Subsequent crystal structures confirmed his insight.

Each binding site takes on a repeating cycle of three conformational states:

O   binding site empty, 'open' state

L   'loose' state, binding ADP and P_i

T   'tight' state, binding ATP

Fig. 4.15 The binding-change scheme of P.D. Boyer for the mechanism of ATPase. The molecule contains three binding sites, which interconvert between three conformational states as the molecule rotates. This diagram shows one stage of the active cycle. The three $\alpha\beta$ dimers have three different states. In 1, the open state O (orange) is empty, the loose state L (green) contains ADP + P$_i$, and the tight state T (blue) contains ATP. In a logical intermediate stage (bracketed), rotation of the γ, δ and ε subunits (not shown) within the $(\alpha\beta)_3$ hexamer converts the L state to a T state, the T state to an O, and the O state to an L. The L state can accept a new charge of substrate. The T state can form ATP. The O state can release ATP.

At stage 2, the ATP has fallen out of the O state, new ADP + P$_i$ have bound to the L state, and ATP has been synthesized in the T state. Note that the bracketed state is presented solely for explanation, and does not represent a physically trappable intermediate. Comparison of states 1 and 2 shows that the molecule has been returned to its initial state but with *different* subunits in the L, O, and T states.

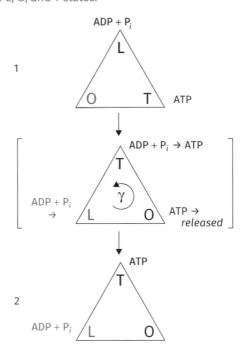

terms of the concentrations of individual ions, but voltage difference provides a summation of the effects of the individual ions.

In the resting state of a neuron, pumps establish concentration disequilibria of Na$^+$ and K$^+$ ions across the axon membrane ([Na$^+$] higher outside; [K$^+$] lower outside). *Other* Na$^+$ and K$^+$ channels in the membrane—not the pump—are closed in the resting state. But opening these channels would release the spontaneous flow of Na$^+$ into the cell and K$^+$ out, to equilibrate the concentrations.

The state of these channels depends on the transmembrane voltage. In the resting state they are closed. The arrival of a neurotransmitter discharged by a pre-synaptic neuron depolarizes the membrane, changing the voltage, and opens first the **voltage-gated sodium channel**, and subsequently the voltage-gated potassium channel. The flow of ions across the membrane of the nerve cell is propagated down the axon. After firing, the pumps restore the concentration disequilibria that characterize the resting state. The neuron will then be poised to fire again.

4.6   Signal reception and transduction

Communication among different parts of our bodies has two available routes: (1) the nervous system, and (2) secretion of a signalling molecule, such as a hormone, at one site, for instance the pituitary gland; and reception of the signal by a cell in one or more distant locations, for instance the adrenals. (Secretion and **signal reception** *can* also be local—in the transmission of a nerve impulse across a synaptic gap, the upstream cell releases neurotransmitter molecules that diffuse to the downstream cell—across a distance of typically 20–40 nanometres between central nervous system neurons, slightly higher at neuromuscular junctions.)

Cells have invented an ingenious mechanism whereby the arrival of a signal at the cell surface can be communicated to the cell interior, without the signalling molecule's needing to enter the cell. The receptor is a transmembrane protein, with domains protruding outside and inside the cell. In the resting state—in the absence of a signalling molecule—the receptor proteins do not interact, and are free to diffuse laterally within the membrane. The binding site for the ligand is created by the extracellular domains of *two* receptor proteins. When a signalling molecule arrives at the cell surface, it causes the two receptor proteins to dimerize (see Figure 4.16). Figure 4.17 shows an actual example: the binding of human growth hormone to its receptor.

The trick is that bringing together the extracellular domains pulls together the intracellular domains also (see Figure 4.16 right)). The intracellular domains have a kinase activity. The two intracellular domains—brought together by the binding of the signal—cross-phosphorylate each other, at selected tyrosine residues. This post-translational modification turns on the enzymatic activity of the intracellular domains. They can catalyse additional intracellular reactions, activating other intracellular proteins, to initiate a signalling cascade.

Fig. 4.16 Dimerization mechanism for transmission of a signal across a cell membrane. Receptor molecules contain exterior, transmembrane, and interior segments. In the absence of a ligand they are monomeric (left). Binding of a ligand brings together the exterior *and* the interior domains. The dimerization of the interior domains activates processes inside the cell. The signalling molecule does not need to enter the cell.

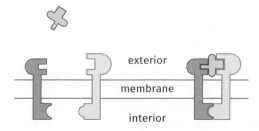

Fig. 4.17 Human growth hormone (blue) in complex with two molecules illustrating the dimerized exterior domain of its receptor (green, orange). This figure shows only the extracellular domains. The transmembrane and intracellular domains do not appear.

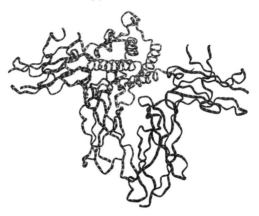

💡 **Key point**

The receptor-dimerization mechanism allows cells to recognize the arrival of a signal at the cell surface, and to trigger responses from intracellular control networks, *without* the signalling molecule's entering the cell.

G-protein-coupled receptors

G-protein-coupled receptors are a large group of cell surface receptors that receive extracellular signals and initate intracellular responses. They have a typical structure containing seven transmembrane helices (see Figure 4.18).

The downstream partners of GPCRs in signal-transduction pathways are heterotrimeric G proteins. These comprise three subunits: Gα, Gβ, and Gγ. Gα and Gγ are anchored to the membrane (see Figure 4.19(a)).

In the resting, inactive state, the G protein is trimeric: G$\alpha\beta\gamma$, in which Gα binds GDP. Activation of a GPCR catalyses GDP–GTP exchange in the Gα subunit. This destabilizes the trimer, dissociating Gα:

$$\text{G}\alpha \,(\text{GTP})\text{G}\beta \,\text{G}\gamma \rightarrow \text{G}\alpha \,(\text{GDP}) + \text{G}\beta \,\text{G}\gamma$$

The two components Gα and GβGγ activate downstream targets, such as adenylate cyclase. This propagates the signal, and initiates complex control cascades. The effects can reach to the nucleus, to modulate gene expression.

G proteins are reset to the resting state via GTPase activity of Gα, converting Gα(GDP) $\rightarrow$ Gα(GDP). Gα(GDP) does not bind to its downstream receptors, shutting down the pathway of signal transmission. Instead, Gα(GDP) rebinds the GβGγ subunits, resetting the system to await the next stimulus.

Fig. 4.18 The seven transmembrane-helices of a G-protein-coupled receptor, or GPCR (the κ opioid receptor). The viewpoint is in the plane of the membrane, and the axes of the helices are approximately perpendicular to the membrane. In this view the extracellular side is up and the cytoplasmic side is down. The helices are arranged—very roughly—in a row of four, in front, coloured magenta; and a row of three, behind, coloured cyan. Many GPCRs contain a short C-terminal helix with an axis in the plane of the membrane.

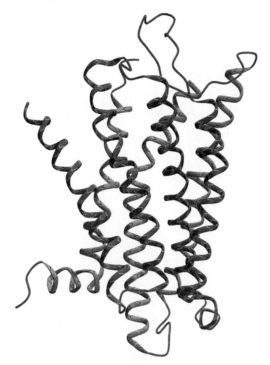

The GTPase activity of Gα is stimulated by a class of proteins called regulators of G-protein signalling (RGSs) (see Figure 4.19(b)).

 Key point

An essential part of any signalling process is a mechanism for cancelling the activation—The Sorcerer's Apprentice offers an example of the consequences of turning a signal on without a mechanism for turning it back off!

4.7 Classification of protein function

A recurrent series of problems arises whenever a new genome is sequenced. To produce useful annotation we need to know:

1. What proteins does the genome encode?
2. What are the functions of these proteins?

Fig. 4.19 (a) A GPCR containing seven transmembrane helices interacts with a heterotrimeric G protein, which contains subunits α, β, and γ. (b) The activation–deactivation cycle of a G protein. Activation catalyses GDP–GTP exchange, resulting in dissociating α from the $\beta\gamma$ subunits. These then activate downstream targets, initiating a control cascade. GTPase activity in the α subunit recovers GDP binding and reassociation of the trimer. This resets the system.

(a)

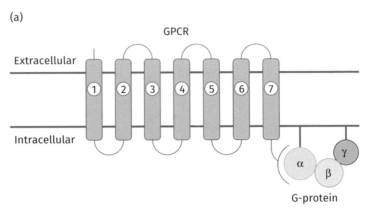

(b)

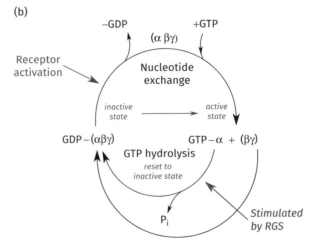

Answers to these questions go a fair way to defining the 'lifestyle'—at least at the molecular level—of the organism.

To determine the proteins inherent in a genome, it is necessary first to identify the protein-coding regions, and then to translate them into amino-acid sequences. Methods for addressing these problems give reasonable answers, although one must recognize that the information about the proteins that the genome contains is incomplete. In particular the genome does not tell us about:

1. three-dimensional structures

2. splice variants—the selection for translation of different combinations of regions of the messenger RNA transcribed from a gene in the DNA

3. cofactors—for example, many proteins contain NAD as a stable component of their structures

4. post-translational modifications

5. oligomer formation—it would be impossible to deduce from the human genome sequence that haemoglobin is a tetramer containing two α and two β chains.

Suppose, however, that we have identified and studied the proteins from some cell. We want to assign their functions. It is useful, for this purpose, to have a catalogue of protein functions. Then the problem becomes: in which 'pigeonhole' does the protein belong?

Two catalogues of protein function are available: (1) The Enzyme Commission classification. This dates from 1961. It covers *only* catalytic functions. Unsurprisingly, it has been extended by the (2) Gene Ontology classification, devised in 2000 in connection with the task of annotating the *Drosophila melanogaster* genome.

 Key point

Catalogues of protein functions allow annotation of function by assignment of each protein to one or more entries in the classification. This is better than asking individual scientists to write descriptions of protein functions. That would make it harder to recognize similarities between different descriptions of the same function.

The Enzyme Commission

International scientific bodies, such as the International Union of Biochemistry (IUB) and the International Union of Pure and Applied Chemistry (IUPAC), are responsible for organizing the intellectual infrastructure of science.

In 1955, the General Assembly of the International Union of Biochemistry, in consultation with the International Union of Pure and Applied Chemistry, established an International Commission on Enzymes, to systematize nomenclature. The Enzyme Commission (EC) published its classification scheme, first on paper and now on the web: https://www.qmul.ac.uk/sbcs/iubmb/enzyme/. This was the first detailed classification of protein functions.

EC numbers (looking suspiciously like computers' IP numbers) contain four fields, corresponding to a four-level hierarchy. For example, EC 1.1.1.1 corresponds to alcohol dehydrogenase, catalysing the general reaction:

an alcohol + NAD$^+$ = the corresponding aldehyde or ketone + NADH + H$^+$

Several reactions, involving different alcohols, would share this number; but the same dehydrogenation of one of these alcohols by an enzyme using the alternative cofactor NADP$^+$ would be assigned EC 1.1.1.2.

The first field in an EC number indicates one of the seven main divisions (classes) to which the enzyme belongs:

Class 1.	Oxidoreductases	catalyse oxidation-reduction reactions
Class 2.	Transferases	transfer of a functional group from one molecule to another
Class 3.	Hydrolases	split a substrate by hydrolysis
Class 4.	Lyases	non-hydrolytic addition or removal of groups from substrates
Class 5.	Isomerases	intramolecular rearrangement
Class 6.	Ligases	join two molecules together
Class 7.	Translocases	control movement of molecules across membranes, or their separation within membranes.

The significance of the second and third fields depends on the class:

- For *Oxidoreductases*, the second number describes the substrate and the third number the acceptor.
- For *Transferases*, the second number describes the class of moiety transferred, and the third number describes either more specifically what they transfer or in some cases the acceptor.
- For *Hydrolases*, the second number signifies the kind of bond cleaved (e.g. an ester bond) and the third number the molecular context (e.g. a carboxylic ester or a thiolester). (Proteinases are treated slightly differently, with the third number indicating the mechanism: serine proteinases, thiol proteinases, and acid proteinases are classified separately.)
- For *Lyases*, the second number signifies the kind of bond formed (e.g. C–C or C–O), and the third number the specific molecular context.
- For *Isomerases*, the second number indicates the type of reaction and the third number the specific class of reaction.
- For *Ligases*, the second number indicates the type of bond formed and the third number the type of molecule in which it appears. For example, EC 6.1 for C–O bonds (enzymes acylating tRNA), EC 6.2 for C–S bonds (acyl–CoA derivatives), etc.
- For *Translocases*, the second number indicates the types of ion or molecule translocated, and the third number the reaction that provides the driving force (where relevant).

The fourth number is the most precise function specification.

The EC produced a catalogue of reactions, not an assignment of function to proteins. The EC has emphasized that: 'It is perhaps worth noting, as it has been a matter of long-standing confusion, that enzyme nomenclature is primarily a matter of naming reactions catalysed, not the structures of the proteins that catalyse them.' (https://www.chem.qmul.ac.uk/sbcs/iubmb/nomenclature/)

Assigning EC numbers to proteins is a separate task. Such assignments appear in protein databases such as UniProt.

Specialized classifications are available for some families of enzymes. For instance, the MEROPS database by N.D. Rawlings and A.J. Barrett provides a structure-based classification of peptidases and proteinases: https://www.ebi.ac.uk/merops/

The Gene Ontology™ Consortium

In 1999, Michael Ashburner and many coworkers faced the problem of annotating the soon-to-be-completed *Drosophila melanogaster* genome. They found the EC classification unsatisfactory, if only because it was limited to enzymes. Ashburner organized the Gene Ontology (GO) Consortium to produce a standardized scheme containing a comprehensive classification of gene function. (An ontology is a formal set of well-defined terms with well-defined interrelationships; that is, a dictionary and rules of syntax.) Ashburner described the project in a memoir: see Further Reading.

As with the EC classification, GO provides a catalogue of functions, not an assignment of the function of particular genes or proteins.

 Key point

The Gene Ontology™ Consortium (http://www.geneontology.org) has produced a systematic classification of gene function, in the form of a dictionary of terms, and their relationships.

Organizing concepts of the Gene Ontology project include three categories:

- *Molecular function*: a function associated with what an individual protein or RNA molecule does in itself; either a general description such as *enzyme*, or a specific one such as *alcohol dehydrogenase*. This is function from the biochemist's point of view.

- *Biological process*: a component of the activities of a living system, mediated by a protein or RNA, possibly in concert with other proteins or RNA molecules; either a general term such as *signal transduction*, or a particular one such as *cyclic AMP synthesis*. This is function from the cell's point of view.

Because many processes are dependent on location. Gene Ontology also tracks:

- *Cellular component*: the assignment of site of activity or partners. This can be a general term such as *nucleus* or a specific one such as *ribosome*.

An example of the GO classification is shown in Figure 4.20. The GO schemes are not strict hierarchies, but have a more general structure.

Fig. 4.20 Selected portions of the three categories of Gene Ontology (GO), showing classifications of functions of proteins that interact with DNA. (a) Molecular function: including general DNA binding by proteins, and enzymatic manipulations of DNA. (b) Biological process: DNA metabolism. (c) Cellular component: different locations within the cell, or partners.

These pictures illustrate the general structure of the GO classification. Each term describing a function is a node in a graph. Each node has one or more parents and one or more descendants: arrows indicate direct parent-child relationships. Unlike the EC hierarchy, the GO graphs are not trees in the technical sense, because there can be more than one path from an ancestor to a descendant. For example, there are two paths in (a) from enzyme to ATP-dependent helicase. Along one path helicase is the intermediate node; along the other path adenosine triphosphatase is the intermediate node.

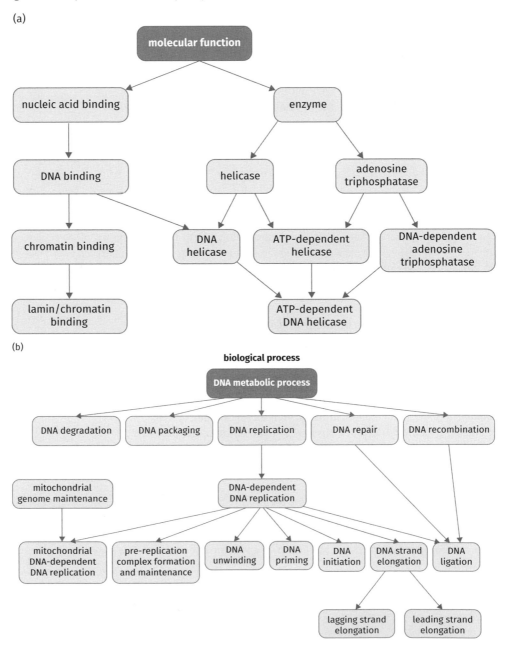

Fig. 4.20 (*Continued*)

(c)

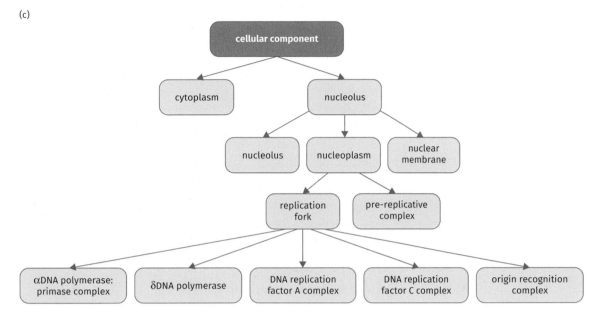

Prediction of protein function

Associated with assignments of GO categories to proteins—https://www.ebi.ac.uk/GOA—are evidence codes of variable reliability, from 'IDA: Inferred from Direct Assay' down to 'IEA: inferred from electronic annotation'.

Indeed, the most common method for assignment of function is by transfer of annotation from a homologue. Calibrations of sequence differences between homologues, and the similarity of EC numbers and GO category assignments, show that for sequence identity >35–50%, in most cases specific function is preserved. For higher degrees of divergence, similarity only in general functional class may be retained. Fetrow and Babbitt comment that: 'Most often, misannotations arise from transfer of a more detailed molecular function than is warranted.'[1]

If one or more relevant structures are known, methods used for drug design are applicable to function prediction also. These include computational prediction of possible ligands, and the sites in the protein at which they might bind.

The **Critical Assessment of protein Function Annotation (CAFA)** programme organizes assessments of blind tests of computational methods for predicting protein function from amino-acid sequence. The goals are prediction of Gene Ontology (GO) terms in all three categories: Molecular Function, Biological Process, and Cellular Component.

[1]Fetrow, J.S. and Babbitt, P.C. (2018). New computational approaches to understanding molecular protein function. PLoS Comput Biol. 14, e1005756.

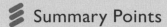 Summary Points

- Proteins provide a wide variety of functions to support the architectures and life processes of viruses, cells, and organisms. Some are structural; some are enzymes—that is, catalysts, and others have regulatory roles.

- A reaction diagram showing the free-energy changes in an enzymatic reaction presents the fundamental explanation of the thermodynamics and kinetics of enzymatic catalysis.

- Many enzymes show similar features in a graph of initial velocity against substrate concentration, in accord with the Michaelis-Menten model of enzymatic catalysis. From this graph it is possible to derive values that characterize enzymatic activity—K_M and V_{max}.

- Immunoglobulins are multidomain proteins sharing a common basic structure. The variety of their binding sites allows them to recognize the organic world. They protect us from toxins and diseases. By genetic engineering their utility in clinical diagnosis and therapy has been very usefully extended.

- The membranes that surround cells and intracellular organelles have as their basic structure a phospholipid bilayer. Control of traffic across membranes requires membrane-embedded proteins that function as channels or pumps. Other membrane-embedded proteins are responsible for recognizing the arrival of a signal at a cell surface. Transduction of the signal into the cell interior, without the need for the signalling molecule itself to enter the cell, can initiate control cascades.

- The steps in the digestion of glucose—glycolysis, the Krebs cycle, and the electron-transport chain and ATP synthase—result ultimately in the capture of energy in the form of ATP. There is a complicated sequence of free-energy changes, involving chemical energy, osmotic energy (creation of a pH gradient across a mitochondrial membrane), and even mechanical energy, in the 'turnstile' mechanism of the microscopic motor ATP synthase, which couples proton flow to production of ATP.

- To assign function to the proteins suggested by the sequences of a novel genome, it is useful to have catalogues of possible protein functions. The Enzyme Commission and the Gene Ontology projects provide classification schemes for protein functions.

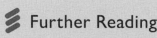 Further Reading

Books and articles

Ashburner, M. (2006). Won for All: How the Drosophila Genome Was Sequenced. (Cold Spring Harbor, NY: Cold Spring Harbor Laboratory Press).
A personal memoir.

Gene Ontology Consortium (2015). Gene Ontology Consortium: Going forward. Nucleic Acids Res. 43 (Database issue), D1049–D1056.
Description of subsequent developments.

Mazumder, R. and Vasudevan, S. (2008). Structure-guided comparative analysis of proteins: principles, tools, and applications for predicting function. PLoS Comput. Biol. 4, e1000.
A detailed guide to assigning function based on mining databases (not laboratory experiments).

Niehaus, T.D., Thamm, A.M.K., de Crécy-Lagard, V. and Hanson, A.D. (2015). Proteins of unknown biochemical function: A persistent problem and a roadmap to help overcome it. Plant Physiol. 169, 1436–42.
Annotation of function by mining databases, plus recommendation for experimental tests with microbial homologues of eukaryotic proteins.

Schnoes, A.M., Brown, S.D., Dodevski, I. and Babbitt, P.C. (2009). Annotation error in public databases: misannotation of molecular function in enzyme superfamilies. PLoS Comput. Biol. 5, e1000605.
A cautionary tale. . .

Stewart, A.G., Sobti, M., Harvey, R.P. and Stock, D. (2013). Rotary ATPases: Models, machine elements and technical specifications. Bioarchitecture 3, 2–12.
Description of the ATPase mechanism.

Whisstock, J.C. and Lesk, A.M. (2003). Prediction of protein function from protein sequence and structure. Quart Rev. Biophys. 36, 307–40.
A useful review, although rather dated.

Zhou, N. et al. (2019). The CAFA challenge reports improved protein function prediction and new functional annotations for hundreds of genes through experimental screens. Genome Biol. 20, 244.
Report on the latest Critical Assessment of protein Function Annotation (CAFA) programme.

Video clips

https://www.youtube.com/watch?v=LQmTKxI4Wn4
Electron transport chain and ATP synthase.
https://www.nature.com/articles/ncomms1693
Supplementary material contains videos about the ATPase mechanism.
https://www.youtube.com/watch?v=-7AQVbrmzFw
Kinesin 'walking': Molecular Motor Struts Like Drunken Sailor.

≋ Discussion Questions

4.1   In some cases, a disease arising from the absence or dysfunction of a protein can be treated by supplying the protein: insulin for diabetes and clotting factors for haemophilia are examples. Some people have a congenital molecular defect that threatens diminished stature. This may arise from inadequate production of active human growth hormone, or—in Laron syndrome—from a defect in the growth hormone receptor. Consider whether the intravenous administration of external human growth hormone (in the former case) or growth hormone receptor (for Laron syndrome) would be expected to produce normal stature.

4.2   Suppose you wanted to create a new protein that had a novel enzymatic activity. Instead of trying to do this *a priori* (see Discussion Question 3 of Chapter 3), you prefer an experimental approach based on antibodies. How would you proceed to elicit antibodies that might be reasonable bases for engineering of the desired catalytic activity?

4.3   **NOTE: Dinitrophenol is toxic and should not be ingested by humans or animals.** Dinitrophenol acts to short-circuit the action of ATP synthase by making the mitochondrial inner membrane permeable to protons. In the presence of dinitrophenol, protons diffuse through the mitochondrial membrane directly, not passing through the ATP synthase complex.

If dinitrophenol were not toxic, would you expect it to function effectively as a 'diet pill' to help obese patients achieve weight loss even if they consumed large quantities of glucose?

5 PROTEIN EVOLUTION

Learning Objectives

- Recognize, given sequences of proteins related by evolution, the mechanisms that have generated the differences. You will be able to predict, from a set of aligned amino-acid sequences, the expected relationships in the three-dimensional structures.
- Be able to compare the deviation of sequence and structure in families of proteins, with the pattern of the divergence of species in classical taxonomic phylogenetic trees.
- Know the protocols for creating unnatural proteins with novel functions, by directed evolution or by protein design.

Evolution of protein structures and functions has generated rich diversity among living things. There are two major mechanisms of protein evolution: (1) Local exploration of sequence space; most simply, one mutation at a time. Given enough time this can—and has—produced proteins with conserved structure and function with virtually no recognizable signal of the relationship left in the amino-acid sequences. (2) Recombination, i.e. 'mixing and matching' of domain assemblies—the functions of the particular domains may or may not be conserved independent of their assembly into full proteins.

Gene frequencies in populations can change either by selection of favourable variants, or nonselective drift. Moreover, many important differences between species of higher organisms are the effect of differences in regulation, rather than the differences in amino-acid sequences of individual proteins.

In addition to evolution in nature, the creation of novel proteins in the laboratory is possible, even the *a priori* design of proteins, with functions unknown in nature.

5.1   Evolution is exploration

Darwin's theory of evolution has two elements: (1) Generation of variation, and (2) Selection of favourable variants by differential reproduction. The application to protein evolution depends on an extended 'Central Dogma' (introduced in Chapter 2):

DNA → RNA → amino-acid sequences of proteins →
protein structures → protein function

Changes in DNA—and to some extent, processing of messenger RNA—generate variation. Selection acts on the proteins expressed. *Selection reinforces innovation.*

Mutations allow proteins to explore the consequences of changes in amino-acid sequences. Some mutations may degrade function; others may completely prevent folding. Such mutations in an essential protein are often lethal. Other mutations may 'improve' function—in the sense that the mutant protein serves a purpose, for the species in which it arises, and under conditions in which a population is currently living, better than the original one. The cell bearing the mutant will have a selective advantage. The mutant gene and protein might replace the original one in the population—natural selection operating at the molecular level. Other mutations may not offer any significant selective advantage or disadvantage. Nevertheless, a population may switch to them by chance, or lose them, a process called *genetic drift*. This is more likely to occur in small populations, such as a limited set of 'founder' individuals colonizing an island.

Proteins can also evolve *within* a species (see The bigger picture 5.1). In 1970, S. Ohno proposed that gene duplication would facilitate development of novel functions. Gene duplication permits separate divergence of the two copies—one copy may continue to provide an essential function; the other may develop a new one. For instance, the related enzymes malate and lactate dehydrogenase catalyse similar reactions, with different substrate specificities:

Malate dehydrogenase:

$$\begin{array}{ccc}
\text{Malate} & & \text{Oxaloacetate} \\
\end{array}$$

$$
\begin{array}{c}
COO^- \\
| \\
CH_2 \\
| \\
HO-C-H \\
| \\
COO^-
\end{array}
\; + \; NAD^+ \; \rightarrow \;
\begin{array}{c}
COO^- \\
| \\
CH_2 \\
| \\
C=O \\
| \\
COO^-
\end{array}
\; + \; NADH \; + \; H^+
$$

Lactate dehydrogenase:

$$\begin{array}{ccc}
\text{Lactate} & & \text{Pyruvate} \\
\end{array}$$

$$
\begin{array}{c}
COO^- \\
| \\
HO-C-H \\
| \\
COO^-
\end{array}
\; + \; NAD^+ \; \rightarrow \;
\begin{array}{c}
COO^- \\
| \\
C=O \\
| \\
COO^-
\end{array}
\; + \; NADH \; + \; H^+
$$

The bigger picture 5.1
Homologues, orthologues, and paralogues

Related proteins are derived from a common ancestor. They are called **homologues**. Corresponding related proteins in *different* species are **orthologues**. The haemoglobins of human, horse, and turkey are orthologues (see Figure 2.12 and 2.13). Homologous proteins in the *same* species are **paralogues**. The α and β chains of human haemoglobin are paralogues. Other globin paralogues in humans include embryonic and foetal haemoglobin, myoglobin, cytoglobin, and neuroglobin.

Discussion Questions

1. In many cases, protein evolution proceeds by the classical *ancestor →
selection → descendant* route. This accounts for the correspondence between divergence of orthologues in a protein family at the molecular level—for instance, in mammalian haemoglobin α chains (see Figures 2.12 and 2.13)—and the standard phylogenetic tree of the species.
   What is the effect on this picture, of horizontal gene transfer? (Horizontal gene transfer is the movement of genetic material between cells or species in ways *other* than the standard parent-to-offspring route; for instance by exchange of plasmids in bacteria or via viral carriers between higher organisms.) How would you expect horizontal gene transfer to disturb the correlation between patterns of divergence at the molecular and species levels?
2. What if the duplication of a gene and the divergence of the two copies is followed by loss of a different copy in two descendant species? Thus suppose the duplication of gene A to A1 and A2 in an ancestor species is followed by loss of A1 in one descendant line and by loss of A2 in another descendant line. Are A1 and A2 orthologues or paralogues? If the original species becomes extinct and you have genome sequences for the two descendant lines, how could you distinguish whether A1 and A2 are orthologues or paralogues?

To study protein evolution, we must first assemble, for proteins of different species, the data on amino-acid sequence, three-dimensional structure, and function. On the basis of these data, it is possible to classify the proteins into homologous families. This must be done at the level of individual domains.

An ultimate goal is to understand *how* differences in sequences, structures, and assembly of domains modify function and even create novel function.

> 💡 **Key point**
>
> For individual domains we can compare the amino-acid sequences and the structures. For complete proteins we can compare the assembly of domains, and the quaternary structure. We can also, both for individual domains and for assemblies of domains, compare the functions of the proteins. In some cases, local mutations—that keep the overall sequence and structure intact but make crucial modifications—can provide novel functions. In other cases, reassembly of domains allows still broader functional horizons.

5.2   The importance of regulation

Still another aspect of protein evolution can occur with maintenance of function, but change in regulation. In some cases this arises with changes in domain composition, or quaternary structure, of proteins.

It is often the case that eukaryotic enzymes have more complex regulatory properties than prokaryotic homologues (see Case study 5.1). For instance, the enzyme 5,10-methylenetetrahydrofolate reductase catalyses the production of 5-methyltetrahydrofolate, needed as a cofactor in the biosynthesis of methionine. Eukaryotic homologues contain an N-terminal catalytic domain and a C-terminal regulatory domain, each approximately 300 residues long. Bacterial homologues lack the regulatory domains (see Figure 5.1).

Also very important are controls over protein expression, such as those that create different distributions of proteins in different tissues—despite the fact that almost all cells in an organism have the same genome sequence.

> 💡 **Key point**
>
> It was pointed out by M.-C. King and A.C. Wilson, and others, that many essential differences between humans and chimpanzees are to be found in regulatory—including but not limited to developmental—programmes, rather than in the differences in amino-acid sequences of the individual proteins encoded in the respective genomes. Approximately 29% of corresponding proteins in human and chimpanzee are identical in amino-acid sequence; and most differ by no more than two amino acids, the implied assumption is one mutation in each lineage.

Mammalian haemoglobins and myoglobins provide another example of variation in regulation at the protein level. Haemoglobin undergoes an allosteric change, between a high-affinity form when the oxygen partial pressure is high, to a low-affinity form when the oxygen partial pressure is low. It has the opposite relative affinities for carbon dioxide. In contrast, myoglobin (to which haemoglobin delivers oxygen) is monomeric. The binding of oxygen to myoglobin is a simple equilibrium, with no variation in oxygen affinity.

Fig. 5.1 Superposition of the catalytic domains of 5,10-methylenetetrahydro-folate reductase from *E. coli* (magenta) and human (cyan). The eukaryotic molecule has an additional C-terminal regulatory domain, not present in prokaryotic homologues. The common domain is binding the cofactor flavin adenine dinucleotide (FAD). The regulatory domain is binding the allosteric effector S-adenosylhomocysteine, interacting with an Asn sidechain.

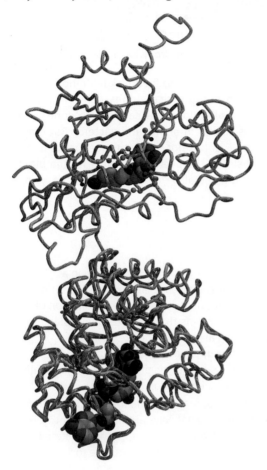

Case study 5.1

Application of regulatory differences among species to genetic engineering: **high-lysine rice**.

Many people in the world subsist primarily or even entirely on rice. Natural rice provides inadequate concentrations of the essential amino acid lysine, with the result that lysine deficiency is widespread in many populations.

Would it be possible to create strains of rice, and other crop plants, that are richer in lysine?

An intermediate in the synthesis of lysine is dihydropyridine-2,6-dicarbo-xylate, synthesized by the enzyme dihydrodipicolinic acid synthase (DHDPS) shown in the following diagram:

Pyruvate   Aspartate semialdehyde   Dihydropyridine-2,6-dicarboxylate

Dihydrodipicolinic acid synthase is subject to feedback inhibition by the product, lysine (see Figure 1.5). To produce rice strains richer in lysine, it would be useful to disable this inhibition. How can this be done? The pathway of lysine biosynthesis is similar in plants and bacteria. But dihy-drodipicolinic acid synthase from *E. coli* is *not* subject to feedback inhibition. Genetic engineering to substitute the bacterial enzyme for the plant enzyme has produced strains of rice, and also maize, with enhanced lysine content.

Discussion Questions

1. Consider a disease in humans that arises from a mutation in a protein. Examples include sickle-cell disease, arising from mutation in haemoglobin, and phenylketonuria, arising from a dysfunction of phenylalanine hydroxylase (the enzyme that converts phenylalanine to tyrosine). In some cases, such as diabetes or haemophilia, a deficient human protein can be supplied directly. One approach to treatment of sickle-cell disease is genetic engineering. It is technically feasible to use CRISPR to 'revert' the mutation.

   In what ways is each of these approaches similar to the production of lysine-rich rice, as described? In what ways are they different? It is currently possible to treat phenylketonuria by supply of a bacterial enzyme, phenylalanine ammonia lyase, that prevents the toxic buildup of phenylalanine by converting it, not to tyrosine, but to trans-cinnamic acid. What problem arises in introducing a bacterial protein in a human patient that would not be a problem in plants?

2. European countries—including, currently, the UK—have imposed restrictions on genetically modified crops, out of concern about dangers to human and animal health, or to the environment.

   The current UK government has called for loosening the restrictions on genetically-modified organisms, as a component of the changes to be introduced by the UK's leaving the EU.

   Discuss the arguments for and against allowing farmers to grow genetically-modified crops. What are the dangers that opponents fear? Are they likely to apply to lysine-enriched rice?

5.3 How do we measure the evolutionary divergence of proteins?

Here is a simple example: lysozymes from human and dog are closely-related proteins. An alignment of their amino-acid sequences (Figure 5.2(a)) shows the conserved residues (54% of the total sequences). A superposition of the structures (Figure 5.2(b)) shows that, despite the substitution of almost half the residues, the course of the mainchain is quite similar. The

Fig. 5.2 (a) Alignment of amino-acid sequences of human and dog lysozyme. The colours reflect the physicochemical type of the residues: yellow = small hydrophobic; green = large hydrophobic; pink = polar, uncharged; red = negatively charged; blue = positively charged. Letters below the bands of human and dog sequences indicate residues conserved between these two proteins. (b) Superposition of tracings of the main-chains of the structures (human = blue; dog = magenta). The boxed region shows a well-conserved 'core' of the structures. (The deletions in the sequences around position 50 correspond to the conformational change in the loop at the top centre of Figure 5.2(b).)

(a)

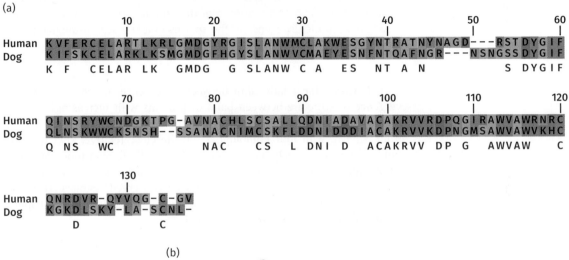

(b)

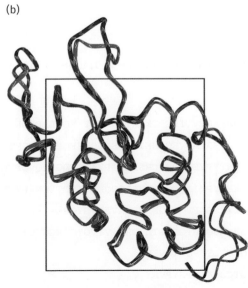

region in the box in Figure 5.2(b) is a well-conserved core of the structure, showing a better fit than the peripheral regions. Both proteins have the function of hydrolysing bacterial cell wall polysaccharides. The binding site is well-preserved in structure and sequence.

Human and dog lysozymes are closely-related proteins. If we comprehensively compare proteins in different species, we find many sets of similar proteins with corresponding functions. There are many homologues of human and dog lysozymes in other mammals. As another example, many species have haemoglobins. Like human and dog lysozymes, human and animal haemoglobins are similar but not identical in amino-acid sequence (see Figure 2.12) and in three-dimensional structure.

 Key point

Many families of proteins show common patterns of divergence in different species. In most cases, the extents of divergence of the DNA gene sequence, the amino-acid sequence of the proteins, and the structure, are congruent to the divergence of the species in a classical taxonomic phylogenetic tree (see Figure 2.13).

To illustrate the relationship between divergence of sequence and divergence of structure between pairs of proteins with increasingly distant relationships, consider comparisons of the sulphydryl proteinase papain from papaya, with four homologues—(a) the close relative, kiwi fruit actinidin, and successively more distant relatives, (b) human procathepsin L, (c) human cathepsin B, and (d) *Staphylococcus aureus* staphopain. In these sequence alignments, vertical lines indicate conserved residues. The similarity in both sequence and structure decreases markedly between papain/actinidin and papain/*S. aureus* staphopain.

(a) Papain / actinidin: total number of positions = 219; number (and %) of identical residues = 101 (46.6%)

(a)
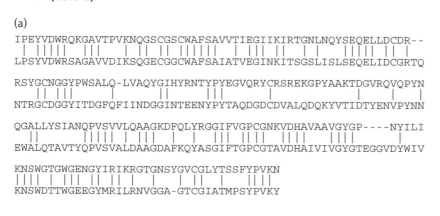

(b) **Papain / human procathepsin L: total number of positions = 220; number (and %) of identical residues = 81 (36.8%)**

(b)

```
IPEYVDWRQKGAVTPVKNQGSCGSCWAFSAVVTIEGIIKIRTGNLNQYSEQELLDCD--R
    ||| || |||||||||| ||| ||||||    ||    ||| |    ||| |||
V----DWREKGYVTPVKNQGQCGSSWAFSATGALEGQMFRKTGRLISLSEQNLVDCSGPE

RSYGCNGGYPWSALQLVAQY-GIHYRNTYPYEGVQRYCRSREKGPYAAKTDGVRQVQPYN
 |||||       | | |   |    ||||    |      |       |   |
GNEGCNGGLMDYAFQYVQDNGGLDSEESYPYEATEESCKYNPKYS-VANDAGFVDIPKQE

QGALLYSIANQPVSVVLQAAGKDFQLYRGGIFVGP--CGNKVDHAVAAVGYG---PNYIL
     |||     | ||    ||     ||   ||    ||| | ||||    |
KALMKAVATVGPISVAIDAGHESFLFYKEGIYFEPDCSSEDMDHGVLVVGYGFESNKYWL

IKNSWGTGWGENGYIRIKRGTGNSYGVCGLYTSSFYPVKN
 |||||  ||   ||  |    ||     ||  ||
VKNSWGEEWGMGGYVKMAKDRRN-H--CGIASAASYPTV-
```

(c) **Papain / human cathepsin B: total number of positions = 251; number (and %) of identical residues = 66 (26.3%)**

(c)

```
IPEYVD-WRQKGAVTPVKNQGSCGSCWAFSAVVTIEGIIKIRTGNLNQYSEQELLD-C-D
      | |        |||||||||| ||   |    ||| |        |   || |
--DAREQWPQCPTIKEIRDQGSCGSCWAFGAVEAISDRICIHTNVSVEVSAEDLLTCCGS

RRSYGCNGGYP------WSALQLVAQYGI--HYRN-TY-----P--YEGVQRYCRSREKG
 ||||||||         |    ||      |    |              |
MCGDGCNGGYPAEAWNFWTRKGLVSGGLYESHVGCRPYSIPPCEHHVNGSRPPCTGEGDT

PYAAK------TDGVRQVQPYNQGALLYSIANQPVSV-V-----LQ---AAGKDFQLYRG
|   |          |       |       |             |       || ||
PKCSKICEPGYSPTYKQDKHYGYNSYSVSNSEKDIMAEIYKNGPVEGAFSVYSDFLLYKS

GIFVGPCGNKV-DHAVAAV--GY--GPNYILIKNSWGTGWGENGYIRIKRGTGNSYGVCG
|     |       ||     |     |  | ||| || ||   |||    |       |
GVYQHVTGEMMGGHAIRILGWGVENGTPYWLVANSWNTDWGDNGFFKILRGQ-DHCGIES

LYTSSFYPVKN
|   |
EVVAGI-PRTD
```

(d) Papain / *S. aureus* staphopain: total number of positions = 219; number (and %) of identical
residues = 25 (11.4%)

(d)

```
IPEYVDWRQKGAVTPVKNQGSCGSCWAFSAVVTIEGIIKIRTGNLNQYSEQELLDCDRRS
                                                    |         |
-----------------------------------------------EQYVNKLENFKIRE

YGCNGGYPWSALQLVAQYGIHYRNTYPYEGVQRYCRSREKG-PYAAKTDGVRQVQPY---
  | |                  | |  | | |         |               |
TQGNNGWCAGYTMSALLNATYNTNKYHAEAVMRFLHPNLQGQQFQFTGLTPREMIYFGQT

--NQGALLYSIANQPVSVVLQAAGKDFQLYRGGIFVGPCGNKVDHAVAAVGYGPNYILIK
    | |          |       |        |       |     || | | |
QGRSPQLLNRMTTYNEVDNLTKNNKGIAIL-GSRVESRNGMHAGHAMAVVGNAKLNNGQE

NSWGTGWGENGYIRIKRGTGNSYGVCGLYTSSFYPVKN-
              | |
VIIIWNPWDNGFMTQDAKNNVIPVSNGDHYQWYSSIYGY
```

As evolving sequences diverge, a point is reached where the results of a
pairwise sequence alignment do not allow unambiguous identification of
the proteins as homologous. It is therefore of the utmost significance that
three-dimensional structure diverges more conservatively than amino-acid
sequence. Despite the low sequence conservation, the similarity of structure
between papain and staphopain in the core regions is nevertheless clear.

 Key point

W.F. Doolittle called the range of amino-acid sequence identity in alignments between 20–30% the **'twilight zone'**. Some pairs of sequences in
this range may be genuine homologues; others may be unrelated. In the
example, the amino-acid sequence identity between papain and *S. aureus*
staphopain is 11.4%—even lower than Doolittle's 'twilight zone'!

5.4   The relationship between divergence of sequence and structure

Comparison of homologous proteins has revealed a general relationship between the divergence of amino-acid sequence and protein structure.

(1) It is possible to distinguish a 'core' of the structures of a family of
proteins from the 'periphery'. The core is a set of major secondary-structural elements, including the active site, that maintain the
same general topology in different related proteins, although the
geometry of their packing differs in detail. (Figure 5.2(b), showing
human and dog lysozymes, and the superpositions of the papain
homologues, illustrate this.)

Fig. 5.3 Relationship between divergence of amino-acid sequence and structure in the cores of families of homologous proteins.

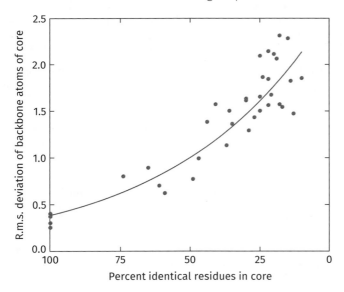

R.m.s. deviation of backbone atoms of core (y-axis, 0.0 to 2.5)

Percent identical residues in core (x-axis, 100 to 0)

Reprinted from Philosophical transactions of the Royal Society of London : Series A, Mathematical and Physical Sciences, 317/56, A.M. Lesk and C. Chothia, The Response of Protein Structures to Amino-Acid Sequence Changes, Copyright (1986), with permission from The Royal Society (UK).

(2) Restricted to the core, there is a relationship between deviation of amino-acid sequence (measured by percent identical residues in the core after optimal sequence alignment) and the deviation of the structures (measured by the average devation of the backbone atoms of the core after optimal superposition).

Different protein families—although completely different in secondary and tertiary structure—fall on a single curve (see Figure 5.3).

The divergence of function is much more complex!

 Key point

Given two sets of corresponding atoms, with coordinates $x_i, y_i, z_i, i = 1, \ldots n$, and $X_i, Y_i, Z_i, i = 1, \ldots n$, to measure structural similarity we compute the **root-mean-square deviation (r.m.s.d.)**: $\sqrt{\sum_{i=1}^{n}[(x_i - X_i)^2 + (y_i - Y_i)^2 + (z_i - Z_i)^2]/n}$ after optimal superposition.

Changes affecting local regions of the genome

Proteins have other routes of evolution than wriggling around in sequence space by successive point mutations. Larger-scale *local* changes in genes include **insertions and deletions** of multiple residues, **inversions**, and **translocations** (see Figure 5.4). We have already mentioned gene duplication.

Fig. 5.4 Possible types of local mutations, larger than single-nucleotide changes, include insertions, deletions, inversions, and copy-number variants. Here red and blue arrows indicate local regions. The combination →---←, common to all diagrams, is an original state; beneath is the result of sequence changes that affect, not a single residue, but a region. What would be the effect of each of these changes on a protein encoded by this region?

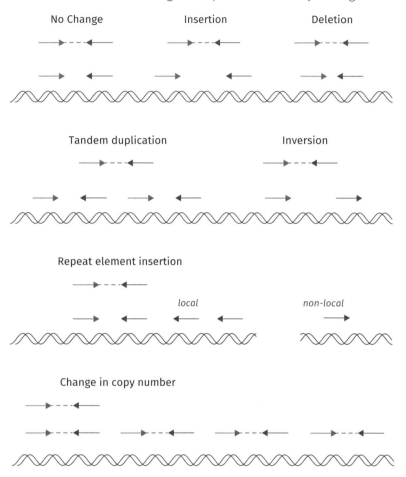

The effects of post-transcriptional events

- A mechanism that breaks the fidelity of mRNA→protein translation is the enzymatic modification of certain bases between transcription and translation, a process known as **mRNA editing**. For instance, in the human intestine, two isoforms of apolipoprotein B appear. To produce one of these forms, mRNA editing alters a single nucleotide, changing a 'c' to 'u'. This turns a Gln codon (caa) into a Stop codon (uaa), truncating the altered isoform to half the size of the longer form. The longer form is the translation of the full unedited mRNA.

- The normal function of ubiquitin is to link to proteins as a signal for their degradation. Ubiquitin+1 (UBB+1) is an aberrant protein arising from deletion of a dinucleotide from mRNA, with 19-amino-acid extension resulting from the frameshift. UBB+1 cannot target proteins for degradation. Even worse, it can accumulate in cells and may contribute to neuronal cell death in Alzheimer's disease.

- Another effect applies at the level of translation. In *E. coli* one gene codes for both the τ and γ subunits of DNA polymerase III. Translation of the entire gene forms the τ subunit. The γ subunit corresponds approximately to the N-terminal two-thirds of the τ subunit. A frameshift on the ribosome leads to premature chain termination 50% of the time, yielding a 1:1 expression ratio of τ and γ subunits. (**Ribosomal frameshifting** is an exception to the processing of mRNA by the ribosome exactly three bases at a time. There is a slippage of the ribosome along the mRNA that changes the reading frame. All organisms, but especially many viruses, use this as a mechanism to synthesize multiple proteins from a single gene. It is not an error but is programmed in the gene.)

Domain reassembly in evolution

A more large-scale, architectural approach to the development of new protein structures and functions is domain recombination. Recall that a domain is a region in a protein that would have independent stability on its own. Given this structural independence, proteins are free to 'mix and match' different combinations of domains (see Figure 2.10). In many cases, variable splicing can contribute to domain recombination. Conversely, mutations in splice sites or termination codons, or deletions in mRNA, can have drastic effects on proteins, almost always deleterious if not fatal.

 Key point

Comparisons of protein sequences and structures confirm that the domain is an important unit of protein evolution. Domains appear in different proteins in different combinations. Thereby, from a relatively small roster of domain families, evolution can assemble a large number of complete proteins.

In some cases, a domain will retain the same function when associated with different partner domains. For instance, many dehydrogenases are multidomain proteins (some multimeric as well) that combine an NAD-binding domain of common structure and function, with a range of partner catalytic domains from seven different families, that vary with the reaction catalysed. The reactions catalysed are coupled to reduction of NAD^+ or $NADP^+$.

In contrast, many examples are known, in which a change in partners, or even a change in domain order along the polypeptide chain, can create or modify catalytic activity. It appears much easier for protein evolution to adapt an existing structure to a new function than to create a new folding pattern. Domain recombination offers great opportunities for evolution of novel functions.

In some cases domains even *change structure* when paired with different partners. As mentioned in Chapter 2, the Nuclear Coactivator Binding Domain NCBD domain of the creb-binding protein is disordered in the unligated state. It forms *different* structures in complex with the Activator for

Thyroid Hormone, and with Retinoid Receptor domain of p160 and with Interferon Regulatory Factor 3.

The enzymes pyruvate decarboxylase and transketolase provide an example of functional change arising from domain reassortment. Pyruvate decarboxylase converts pyruvate to acetaldehyde. Transketolase takes a ketose sugar and an aldose sugar, and converts the ketose to an aldose and the aldose to a ketose. Both enzymes use the cofactor thiamine pyrophosphate.

Structurally, both pyruvate decarboxylase and transketolase contain three domains (see Figure 5.5). They share two of the three domains, but the domains appear in different orders along the polypeptide chains. Nevertheless, the interface between the PYR and PP domains is preserved between the two structures (see Figure 5.6). The active site is formed from residues in these two domains.

Key point

Proteins fairly freely 'mix and match', and reassemble domains, with a variety of functional consequences:

- In some cases, such as the NAD-binding domains of dehydrogenases, the domain retains its function when associated with different partner domains.

- In other cases, entirely novel function emerges.

- In the case of pyruvate decarboxylase, substantial domain rearrangement within the sequence is consistent with retention of a cofactor-binding site in the structure.

Fig. 5.5 (a) Domain architecture of pyruvate decarboxylase (left), comprising PYR = pyrimidine ring binding domain (blue), TH3 = transhydrogenase dIII subunit (magenta), and PP = diphosphate binding domain (green); and (b) transketolase (right), comprising PP (green), PYR (blue), and TKC = transketolase C-terminal domain (orange). The cofactor thiamine pyrophosphate is shown in a shaded-sphere representation.

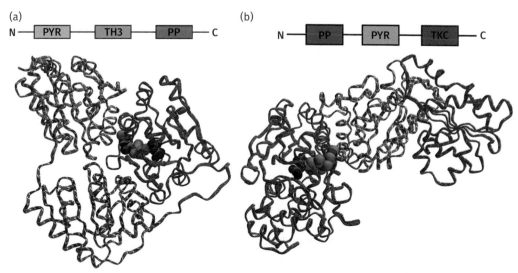

Fig. 5.6 Pyruvate decarboxylase and transketolase, superposed on PYR and PP domains. Despite the difference in overall domain architecture in these two proteins, the geometric relationship between these two domains is preserved. The colours of the domains of pyruvate decarboxylase are the same as in the preceding figure: PYR blue, TH3 magenta, PP green. But in this figure the transketolase domains appear thus: PP red, PYR purple, TKC remains orange. The colours of the PP and PYR domains of transketolase have been changed in order to distinguish them from the superposed domains of pyruvate decarboxylase.

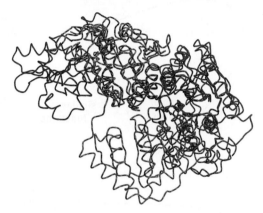

5.5   Pathways and limits in the divergence of sequence, structure, and function

The relationships among sequence, structure, and function are even more complex than the examples in the previous section suggest (see Figure 5.7). Indeed:

- Similar sequences can be relied on to produce similar protein structures, with divergence in structure increasing progressively with the divergence in sequence (see Figure 5.3).

- Conversely, similar structures are often found with very different sequences (see * in Figure 5.7). In many cases the relationships in a family of proteins can be detected only in the structures, the sequences having diverged beyond the point of our being able to detect the underlying common features. (Papaya papain and *Staphylococcus aureus* staphopain are an example.)

- Similar sequences and structures often produce proteins with similar functions, but exceptions abound.

- Conversely, similar functions are often carried out by non-homologous proteins with dissimilar structures. Examples include the many different families of proteinases, sugar kinases, and amino-acyl-tRNA synthetases.

In fact, we must abandon as simplistic the view that sequence uniquely determines structure and that structure uniquely determines function. (1) Many enzymes have multiple functions (they are said to 'moonlight'), or

Fig. 5.7 Relations between similarity of sequence, structure, and function. Black circles indicate sets of sequences of related proteins retaining some degree of similarity in their amino-acid sequences. These correspond to sets of similar structures, in the surrounding red circles. The red circles are larger, because structures can retain similarity even as sequence similarity becomes exiguous. But function—blue ellipses—is not localizable; proteins unrelated in sequence and structure can carry out similar functions.

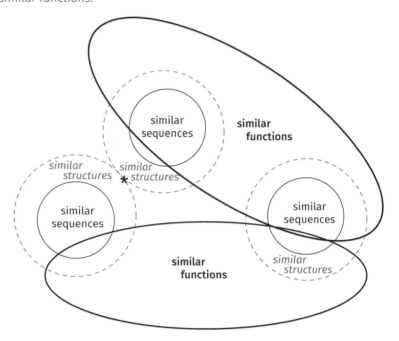

are more or less promiscuous in their substrate or even catalytic specificities; and (2) many enzymes have flexible or even disordered regions. These features are sometimes, but not necessarily, concomitant.

Even if sequence divergence gives rise fairly smoothly to progressive structure divergence, tiny structural changes in sensitive regions can have profound effects on function (see Figure 5.8). Such regions obviously include active sites, but are not necessarily limited to them. It is possible for function to make discrete jumps with minimal sequence change.

 Key point

During evolution it is possible for different branches of a protein family to invent a new function independently, or for a function to be lost and rediscovered. In such cases, similarity of function does not imply close similarity of sequence. The evolutionary tree of the sequences of members of the family, and the graph of the relationships between different functions, can show very different topologies.

Fig. 5.8 From an original structure shown at the centre of this figure, mechanisms of functional change in protein evolution may include one or more of the following: (a) Mutations in residues around the active site, (b) Mutations in residues at interfaces, (c) Gene fusion, (d) Oligomerization (changes in quaternary structure), (e) Promiscuity (= relaxation of specificity), (f) Moonlighting (adopting an additional function), (g) Post-translational modification, and (h) Changes in active site residue. For mechanisms (a), (c), and (d), there are contributions to the new active site from a partner domain or subunit.

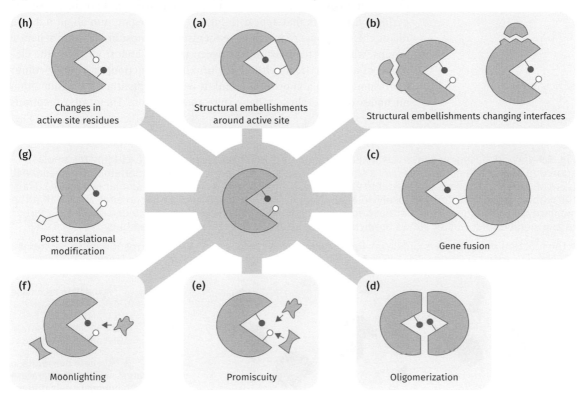

(h) Changes in active site residues

(a) Structural embellishments around active site

(b) Structural embellishments changing interfaces

(g) Post translational modification

(c) Gene fusion

(f) Moonlighting

(e) Promiscuity

(d) Oligomerization

(From: Das, S., Dawson, N.L. and Orengo, C.A. (2015). Diversity in protein domain superfamilies. Curr. Opin. Gen. & Dev. 35, 40–9.)

Divergence of function in the enolase superfamily

The **enolase superfamily** shows wide functional divergence. Enolase itself catalyses the isomerization of 2-phosphoglycerate and phosphoenolpyruvate in glycolysis; however, there are large numbers of homologues, showing at least 20 different catalytic activities.

Some enolase superfamily enzymes that catalyse *different* reactions are more similar to each other, than to other pairs of enzymes that catalyse the *same* reaction. (These might appear in the region in Figure 5.7 where the two blue ellipses overlap.) Conversely, if the activities were invented more than once, independently, some proteins catalysing a common reaction may in fact be distant rather than close relatives. (Indeed, they may not be related at all—there are many evolutionarily-unrelated families of proteases, for example.) These discrepancies frustrate attempts to classify the functions of enzymes on the basis of sequence or even of structure, or to assign functions to all the sequences that have never set foot in a wet lab.

Gerlt and Babbitt, and their coworkers, have described the relationships of the known proteins of the enolase superfamily, in terms of a network (see Figure 5.9). Neighbourhoods in the network depend on sequence similarity. Closely-related proteins form clusters.

Absent from the figure are the enolases themselves, which would loom overwhelmingly large, because so many sequences are known. Those superfamily members that appear in Figure 5.9 segregate into about a dozen major clusters; some of which have themselves almost broken into separate groups, with only tenuous links between them. In almost all cases, the elements of a cluster are all either grey (unknown function) or a single colour (corresponding to a known function). It would be justifiable to assume that all nodes of these clusters have the same function. There are exceptions

Fig. 5.9 Network based on sequence similarity for enolase superfamily members, excluding the enolase subgroup itself. Each node, shown as a square, corresponds to a protein in the superfamily. Lines connect pairs of nodes which have high sequence similarity; shorter edges indicate more-similar sequences. Thus the graph contains clusters of closely-related proteins. The threshold for drawing an edge is such that the similarity of the sequences are within homology-modelling range. Therefore knowledge of the structure of any member of a cluster allows prediction of the structures of other cluster members.

If the function of a protein is unknown, the corresponding node is grey. Nodes corresponding to proteins of known function are coloured. A list of known functions appears at the right of the figure.

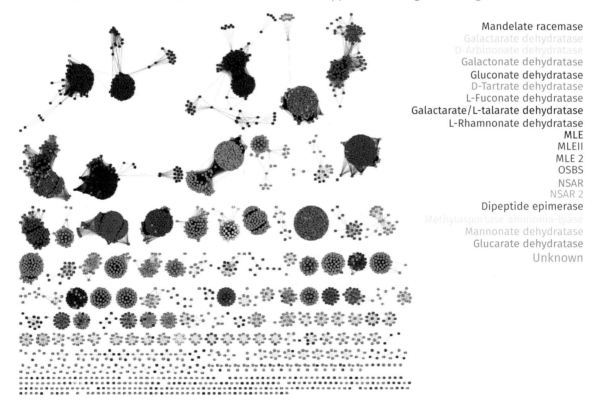

Mandelate racemase
Galactarate dehydratase
D-Arbinonate dehydratase
Galactonate dehydratase
Gluconate dehydratase
D-Tartrate dehydratase
L-Fuconate dehydratase
Galactarate/L-talarate dehydratase
L-Rhamnonate dehydratase
MLE
MLEII
MLE 2
OSBS
NSAR
NSAR 2
Dipeptide epimerase
Methylaspartase ammonia-lyase
Mannonate dehydratase
Glucarate dehydratase
Unknown

(Figure by J.A. Gerlt, based on Gerlt, J.A., Babbitt, P.C, Jacobson, M.P. and Almo, S.C. (2011). Divergent evolution in enolase superfamily: strategies for assigning functions. J. Biol. Chem. 287, 29–34.)

however! For the clusters that are entirely grey, no function of any member of the cluster is known.

Were the enolases themselves added, this figure would display a comprehensive view of the relationships between sequence and function in the superfamily. It would also provide a guide to focus future investigations on interesting and significant questions. A cluster for which no member has a known structure (all nodes grey), would be a good candidate for investigation. If a cluster has the unusual property of containing members with different functions, it would be interesting to see how these minimal sequence changes alter function. If two (or more) separated clusters both show a common function, it may be that this function was invented twice (or more times), independently, in different lineages; it would be interesting to sort this out.

To allow comprehensive and facile access to the data and the analytic tools, Babbitt, Holliday, and coworkers have developed the Structure-Function Linkage Database (SFLD), at http://sfld.rbvi.ucsf.edu/. The data are organized around the relationships between sequences and structural features, and reaction mechanisms. The web site supports browsing, information retrieval, following links to other databases, and interactive analysis, such as computation of networks such as the one appearing in Figure 5.9. Doing this interactively allows selecting different colour schemes, or clicking on nodes to retrieve annotation information for particular molecules. For instance, an operation that would be impossible on the fixed picture in Figure 5.9 but could be done using the interactive tools, would be to identify within the network the particular molecule that corresponds to a node, and determine its most similar homologues. The site is also rich in documentation and tutorial materials, including introductory videos.

5.6   Protein evolution on the lab bench

So far, we have discussed protein evolution in nature. Creation of novel, artificial proteins in the laboratory has succeeded. There are two approaches: (1) directed evolution, and (2) computational protein design.

Directed evolution

One strand of Darwin's thinking that contributed to his theory of evolution was the observation that farmers could improve qualities of livestock by selective breeding. He drew an analogy between this artificial selection and the idea of natural selection that he was proposing as the mechanism of evolution. We now recognize that evolution by natural selection takes place at the molecular level. Why not artificial selection also?

Natural proteins do many things, but not everything we'd like them to. For applications in technology, it would be useful to have proteins that would:

- have activities unknown in Nature;
- show activity towards unnatural substrates or altered specificity profiles;

- be more robust than natural proteins, retaining their activity at higher temperature or in organic solvents;
- show different regulatory responses, enhanced expression, or reduced turnover.

Scientists have used directed evolution—or artificial selection—to generate molecules with novel properties starting from natural proteins.

Evolution requires the generation of variants, and differential propagation of those with favourable features. Molecular biologists dealing with microbial evolution have advantages over the farmers that Darwin observed. We can generate large numbers of variants artificially. Screening and selection can in many cases be done efficiently, by stringent growth conditions. And there are virtually no limits on the size of the 'flock' or 'litter'[1] (but see The bigger picture 5.2).

---

[1]Darwin might well have been envious. He wrote:

> . . . as variations manifestly useful or pleasing to man appear only occasionally, the chance of their appearance will be much increased by a large number of individuals being kept. Hence, number is of the highest importance for success. On this principle Marshall formerly remarked, with respect to the sheep of parts of Yorkshire, 'as they generally belong to poor people, and are mostly in small lots, they never can be improved.'

(Origin of Species, Chapter 1.)

The bigger picture 5.2
We can explore only a tiny fraction of possible sequences

Neither natural nor directed evolution can possibly explore more than a very small fraction of possible polypeptide sequences. The number of amino-acid sequences of N residues is 20^N. To give such numbers some tangible meaning, the total mass of one copy of all possible sequences even of small 18-amino acid polypeptides is larger than the mass of the Earth. The numbers increase very steeply with sequence length.

Discussion Questions

1. The genome of HIV-1, the virus responsible for AIDS, is an RNA molecule 9749 nucleotides long. It is estimated that in an untreated AIDS patient, viral replication will produce double mutants at every pair of positions, every day. Approximately how many different molecules will appear in a patient after one day? After one week? After one month? To what fractions of the total number of possible 9749-long RNA molecules will these amount? (Treat these questions as 'back-of-the-envelope' calculations; do not bother to correct for duplications, and ignore the fact

that many mutations in protein-coding regions will be silent mutations, and many others will be lethal.)

2. The average relative molecular mass of a peptide is 110. What would be an average molecular weight of a 50-residue polypeptide? How many different 50-residue polypeptides are possible? Compare the total mass of a single copy of every possible 50-residue polypeptide, with the mass of the Milky Way (approximately 3×10^{42}g).

The procedure of directed molecular evolution comprises these steps:

1. Create variant genes by mutagenesis or genetic recombination.

2. Create a library of variants by transfecting the genes into individual bacterial cells.

3. Grow colonies from the cells, and screen for desirable properties.

4. Isolate the genes from the selected colonies, and use them as input to step 1 of the next cycle.

Strategies for generating variants include: (1) single and multiple amino acid substitutions, (2) recombination, and (3) formation of chimaeric molecules by mixing and matching segments from several homologous proteins. Each method has its advantages and disadvantages. The smaller the change in sequence, the more likely that the result will be functional. On the other hand, multiple substitutions or recombinations give a greater chance of generating novel features. The choice depends in part on the nature of the goal. For instance, it is easier to lose a function than to gain one. (Why would you want to lose a function? Removal of product inhibition to enhance throughput in an enzymatically catalysed process is an example. See Case study 5.1.)

Directed evolution of subtilisin E

Subtilisins are a family of bacterial proteolytic enzymes. Subtilisin E, from the mesophilic bacterium *Bacillus subtilis,* is a 275-residue monomer. It becomes inactive within minutes at 60°C. Directed evolution has produced interesting variants, with features including enhanced thermal stability, and activity in organic solvents.

•*Enhancement of thermal stability* Thermitase, a subtilisin homologue from the bacterium *Thermoactinomyces vulgaris,* remains stable up to 80°C. The existence of thermitase is reassuring because it provides 'proof of principle' that evolution of subtilisin to a thermostable protein is possible. On the other hand, subtilisin E and thermitase differ in 157 amino acids, more than half. Do we have to go this far? Are all the changes essential for thermostability, or has there been considerable neutral drift as well?

A thermostable variant of subtilisin E, produced by directed evolution, differs from the wild type by only eight amino acid substitutions. The variant is identical to thermitase in its temperature of optimum activity of 76°C (17°C higher than the original molecule) and retains stability at 83°C.

Fig. 5.10 Directed evolution of a thermostable subtilisin. The starting, wild type (wt) was subtilisin E from the mesophilic bacterium *B. subtilis*. Steps of random mutagenesis alternated with recombination. At each step screening for improved properties and artificial selection helped choose candidates for the next round. The ordinate in this graph is the half-life of retention of activity at 65° C.

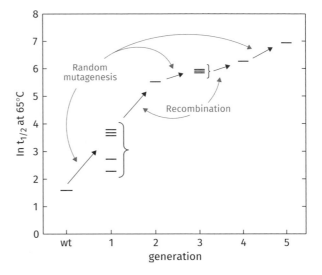

(After: Zhao, H. and Arnold, F.H. (1999). Directed evolution converts subtilisin E into a functional equivalent of thermitase. Protein Eng., 12, 4753.)

The procedure involved successive rounds of generation of variants, and screening and selection of those showing favourable properties (see Figure 5.10). At each step several thousand clones were screened for activity and thermostability.

The optimal variant differed from the wild type at eight positions: P14L, N76D, N118S, S161C, G166R, N181D, S194P, and N218S. The notation, for example P14L, means the amino acid proline (P) at position 14 in the original protein, is mutated to leucine (L) in the final one. Figure 5.11 shows their distribution in the structure. Most of the substitutions are far from the active site, which is not surprising as the wild type and variant do not differ in function. (In this orientation, the catalytic residues are shown in ball-and-stick representation, to the right of the label N218S.) Most of the sites of substitution are in loops between regions of secondary structure. These regions are the most variable in the natural evolution of the subtilisin family. However, only two of the substitutions produce the amino acids that appear, in those positions, in thermitase. Two of the substitutions are in α–helices, including replacement of proline by leucine at residue 14. This has a certain logic: proline tends to destabilize an α–helix because it costs a hydrogen bond.

• *Activity in organic solvents* For applications to industrial processes, it is useful to have enzymes that work under conditions other than those prevailing *in vivo*. Directed evolution produced another variant of subtilisin E

Fig. 5.11 The sites of mutation in *B. subtilis* subtilisin E that produced a thermostable variant by directed evolution. Sidechains shown are those of the final result.

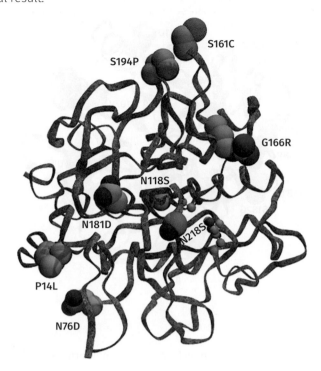

that is active in an organic solvent, 60% dimethylformamide. In this case, no natural homologue with the desired properties was known.

By random mutagenesis and screening, 12 amino acid substitutions were identified that produced a protein with 471 times the proteolytic activity in the organic medium as the wild type, measured in terms of the specificity constant k_{cat}/K_M. The changes were: D60N, D97G, Q103R, I107V, N218S, G131D, E156G, N181S, S188P, Q206L, S188P, and T255A.[2] Note that the only common position affected in both sets is N181.

In this case, eight of the 12 mutations cluster around the active site (see Figure 5.12).

 Key point

Directed evolution of subtilisin has succeeded in producing unnatural 'homologues' active (1) at high temperatures, and (2) in organic solvents.

[2]Chen, K. and Arnold, F.H. (1993). Tuning the activity of an enzyme for unusual environments: sequential random mutagenesis of subtilisin E for catalysis in dimethylformamide. Proc. Nat'l Acad. Sci. USA 90, 5618–22.

Fig. 5.12 The sites of mutation in *B. subtilis* subtilisin E that produced a variant active in organic solvent. Sidechains shown are those of the final result.

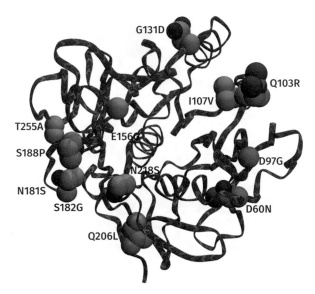

Computational protein design

It is also possible to design novel proteins computationally. Consider the three related problems:

1. *Structure prediction*: Presented with an amino-acid sequence, predict the structure into which it folds.

2. *The 'inverse folding problem'*: Presented with a main-chain conformation, devise an amino-acid sequence that will fold into a structure with that main-chain conformation.

3. de novo *protein design*: Devise an amino-acid sequence that will fold up into a structure that has some desired property, such as a novel enzymatic activity.

Extension of methods for protein structure prediction has achieved solutions of these problems. For example, enzymes designed computationally catalyse the retro-aldol reaction, a Kemp-elimination reaction, and the Diels-Alder condensation (see Figure 5.13). As with directed evolution, it has been possible to create enzymes with catalytic activities unknown in living cells.

De novo protein design can even go beyond that which is accessible from sequences of the 20 natural amino acids. The canonical set of 20 amino acids can be expanded. In the computer, this is trivial. In the laboratory, one could consider partial or even total chemical synthesis, or apply the methods of P.G. Schultz and coworkers that expand the genetic code.

Schultz's group created:

1. novel tRNAs that interact with one of the STOP codons in mRNA (of course, except for suppressor strains, there is normally no tRNA with an anticodon complementary to a STOP codon), and

Fig. 5.13 Reactions catalysed by computationally-designed enzymes: (a) retroaldol reaction. (b) Diels-Alder condensation. (c) Kemp elimination reaction—here B is a general base, provided in the enzyme by a glutamate sidechain.

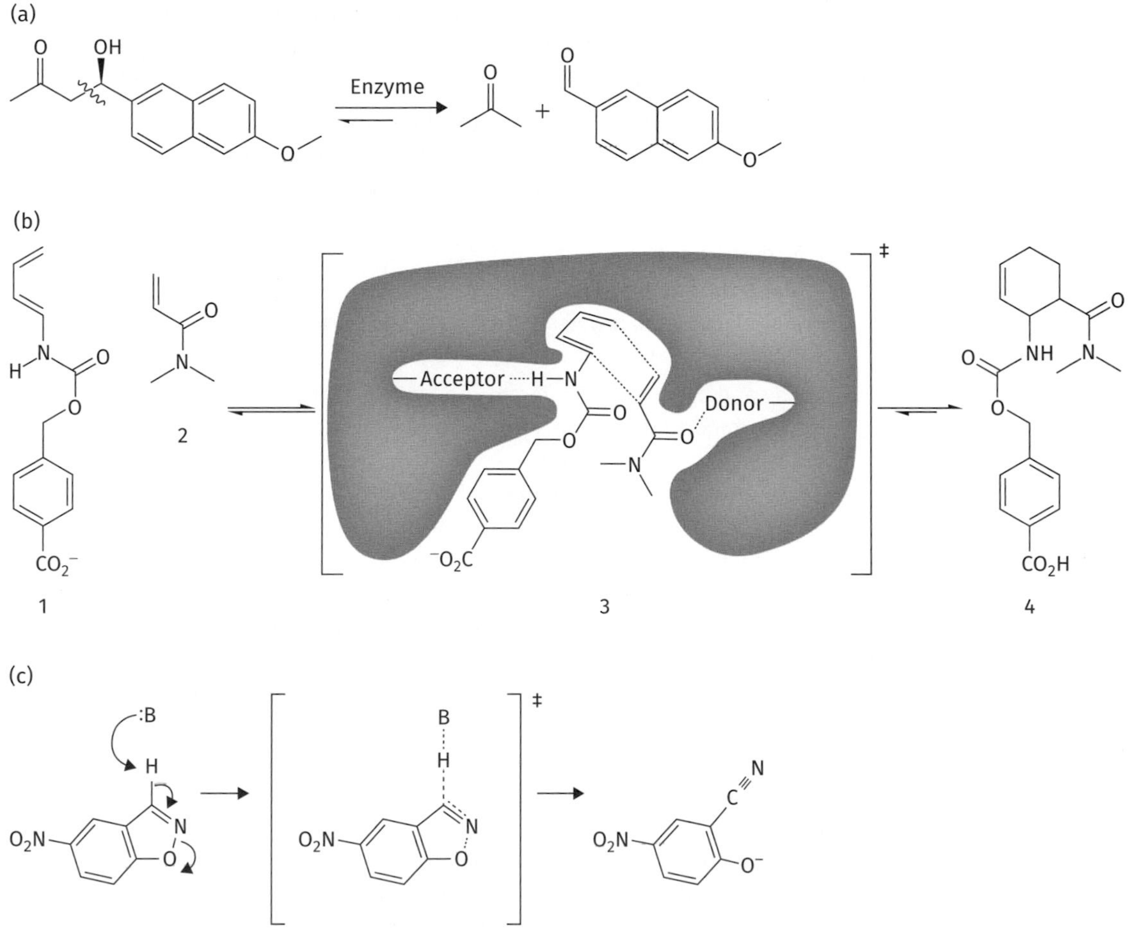

(a)

(b)

(c)

((b) Reprinted from Proceedings of the National Academy of Sciences (PNAS), 109/10, Heidi K. Privett, Gert Kiss, Toni M. Lee, Rebecca Blomberg, Roberto A. Chica, Leonard M. Thomas, Donald Hilvert, Kendall N. Houk, and Stephen L. Mayo, Iterative approach to computational enzyme design, Page 3791, Copyright (2012), with permission from PNAS. (c) From: Siegel, J.B., Zanghellini, A., Lovick, H.M., Kiss, G., Lambert, A.R., St Clair, J.L., Gallaher, J.L., Hilvert, D., Gelb, M.H., Stoddard, B.L., Houk, K.N., Michael, F.E. and Baker, D. (2010). Computational design of an enzyme catalyst for a stereoselective bi-molecular Diels-Alder reaction. Science 329, 309–13.)

2. novel amino-acyl tRNA synthetases that would charge the novel tRNAs with some selected unnatural amino acid.

By inserting the chosen redefined STOP codon at specific positions in a DNA coding sequence (and using other STOP codons to signal chain termination) an unnatural amino acid can be selectively incorporated into a protein synthesized by a ribosome. Schultz's group has devised tRNA-synthetase combinations for over 200 unnatural amino acids. It is even possible to reassign two of the three STOP codons, to insert multiple unnatural amino acids into single proteins.

⬚ Summary Points

- Protein evolution proceeds through generation of variation by mutations in DNA—gene frequencies in populations can change either by selection of favourable variants, or non-selective drift.
- Two components of protein evolution are:
  - **(1)** Divergence of amino-acid sequences within domains.

    Evolutionarily-related proteins in different species form protein families, descended from a common ancestor, called *homologues*. Deviation of both sequence and structure tend to follow the same pattern as the divergence of species in classical taxonomic phylogenetic trees.

    Gene duplication followed by divergence can create related proteins *within* a species, called paralogues.
  - **(2)** 'Mixing and matching' of different combinations of domains is another important mechanism of divergence of protein structure and function.

    This includes: assembly of different complexes of domains within a single polypeptide chain, and, also, changes in the composition or assembly of complexes, i.e. changes in quaternary structure.
- Gene duplication facilitates divergence, as one copy can continue to provide an essential function whilst the other can develop a novel function.
- In addition to changes in amino-acid sequences of individual proteins, important differences between species depend on changes in regulation, either at the protein level or in control of transcription.
- Distinguish the *core* of a family of homologous proteins—a set of major secondary-structural elements, including the active site, that maintain the same general topology, although the geometry of their packing differs in detail—from peripheral regions.
- To compare homologous domains: Similarities of the amino-acid sequences are expressed by the % identical residues in the core in an optimal sequence alignment. Similarities in the structures are expressed by the average distance between corresponding $C\alpha$ atoms in the core after optimal superposition. There is a general relationship between the divergence of the sequence and structure in families of proteins.
- It is possible to create unnatural proteins, with novel functions, in the laboratory.

⬚ Further Reading

Fitch, W.M. (2000). Homology: a personal view on some of the problems. Trends Genet. 16, 227–31.
Valuable insights from one of the major contributors to the field.

Abroi, A. and Gough, J. (2011). Are viruses a source of new protein folds for organisms? – Virosphere structure space and evolution. Bioessays 33, 626–35.
Discussion of the role of viruses in the protein universe.

Das, S., Dawson, N.L. and Orengo, C.A. (2015). Diversity in protein domain superfamilies. Curr. Opin. Gen. & Dev. 35, 40–9.
Evolution at the domain level.

Gerlt, J.A., Babbitt, P.C., Jacobson, M.P. and Almo, S.C. (2012). Divergent evolution in enolase superfamily: strategies for assigning functions. J. Biol. Chem. 287, 29–34.
Background to the section—Divergence of function in the enolase superfamily.

King, M.-C. and Wilson, A.C. (1975). Evolution at two levels in humans and chimpanzees. Science 188, 107–16.
A classic seminal paper.

Baker, D. (2019). What has de novo protein design taught us about protein folding and biophysics? Protein Science 28, 678–83.
Comments by one of the leaders in the field of protein structure prediction and design.

von Heijne, G. (2018). Protein evolution and design. Ann. Rev. Biochem. 87, 101–3; and other articles in this volume.
The above is the introductory chapter to a collection of articles on this theme.

Discussion Questions

5.1 For papain and its homologues (discussed in text), the fraction of identical residues in the core, and the r.m.s. deviation of the Cα atoms of the core, are:

Homologue pair	% identical residues in core	r.m.s. deviation of Cα atoms of core
Papain / actinidin	46.6%	0.68Å
Papain / human procathepsin L	36.8%	0.90Å
Papain / human cathepsin B	36.8%	1.29Å
Papain / *S. aureus* staphopain	11.4%	1.99Å

Plot these points on a copy of Figure 5.3 and discuss whether they fit well.

5.2 In oxygen transport from the lungs to the muscles, haemoglobin absorbs oxygen in the lungs, carries it around the bloodstream, and delivers it to myoglobin in the muscles.

Myoglobin is a monomer. For both myoglobin and haemoglobin, the fraction of molecules binding oxygen does increase with increasing oxygen partial pressure (Le Chatelier's principle). However, for myoglobin the *oxygen affinity* is independent of the oxygen partial pressure (pO_2); that is, oxygen binding to myoglobin (Mb) shows a simple equilibrium:

$$Mb + O_2 = MbO_2, \qquad \frac{[MbO_2]}{[Mb] \cdot pO_2} = K,$$

where the equilibrium constant K is *independent* of oxygen partial pressure.

Haemoglobin is a tetramer, of two α subunits and two β subunits. It undergoes an allosteric change such that the oxygen affinity in the lungs is high—allowing for efficient oxygen capture—but the oxygen affinity in the muscles is low—allowing for efficient oxygen delivery. The allosteric change involves coupled changes of tertiary and quaternary structure. Oxygen binding to haemoglobin cannot be described as a simple equilibrium, as for myoglobin (see Figure 5.14).

Would you expect isolated individual subunits of haemoglobin, or isolated $\alpha\beta$ dimers, to show the unusual oxygen-binding characteristics of the tetramer, or would you expect them to be more like myoglobin? Explain your reasoning.

Fig. 5.14 Oxygen-dissociation curves for myoglobin and haemoglobin. Myoglobin shows a simple equilibrium, with a binding constant independent of oxygen concentration. For haemoglobin, the binding constant for the first oxygen is two orders of magnitude smaller than the binding constant for the fourth oxygen. The units of mmHg for partial pressure are traditional in the literature about this topic. 760mmHg = 1atm = 101,325Pa.

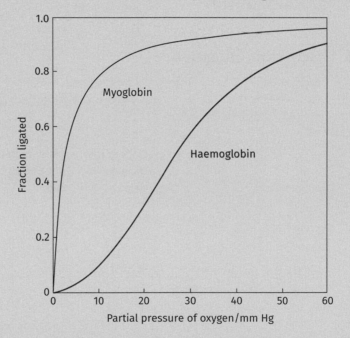

GLOSSARY

α–helix one of the secondary structures common in proteins

α–helix hairpin a supersecondary structure containing two α–helices, oriented in opposite directions and connected by a short loop

α–keratin a fibrous protein forming a common structural protein in our bodies

β–α–β unit a supersecondary structure containing two strands of β–sheet with an α–helix between them

β–hairpin two strands of β–sheet, oriented in opposite directions and connected by a short loop

β–sheet one of the secondary structures common in proteins, comprising several extended strands with lateral hydrogen bonding between neighbouring strands

A

active sites a region of an enzyme, which binds substrates and cofactors, that is the site of the catalysed reaction

affinity chromatography a method of purification that depends on specific binding of a component in a mixture to material in a column

aggregation large-scale association of proteins, often leading to precipitation and disease

alignment arrangement of residues in two or more proteins showing the optimal similarity among the sequences

allosteric changes alteration in a protein conformation, affecting function, caused by binding of ligands away from the active site

AlphaFold2 a computer program based on artificial intelligence that has been very successful in prediction of protein structure from amino-acid sequence

amino acids the components of proteins; there are 20 (occasionally 22) amino acids encoded by triplets of bases in DNA

amino-acid sequence the order of appearance of amino acids in a polypeptide chain

ammonium sulphate precipitation a method of separation of proteins in a mixture, taking advantage of different thresholds of solubility depending on salt concentration

amyloid fibrils a form of aggregated proteins responsible for many diseases

angles of internal rotation definition of a conformation in terms of rotations around single bonds

antibodies a family of molecules in vertebrates that protect against foreign proteins and pathogens

a priori **prediction** prediction of protein structure from amino-acid sequence without making explicit use of known structures

AIDS acquired immune deficiency syndrome

ATP synthase a small molecular motor, embedded in membranes in mitochondria and chloroplasts, that synthesizes ATP

B

Bragg's law expression of condition for constructive interference of X-rays scattered from crystals: $n\lambda = 2d \sin \theta$ (n = some integer, λ = wavelength of the X-rays, d = spacing between planes of atoms in the crystal, θ = angle between the incident (and scattered) X-rays and planes of atoms in the crystal)

C

Central Dogma DNA makes RNA makes protein (Francis Crick, 1957)

chaperone a protein that unfolds misfolded proteins, affording them another chance to fold properly

chemical shift change in nuclear energy levels produced by interactions with the surrounding molecular structure

codon three successive bases in DNA or RNA that encode a specific amino acid, or to signal STOP of translation

cofactor a small organic molecule that is a stable component of a protein structure and participates in function

collagen a major structural protein

competitive inhibitor a molecule that slows down the action of an enzyme by binding to the active site

complementarity-determining regions (CDRs) regions in antibody molecules that control the specificity of binding

computational protein design use of computer programs to predict unnatural amino-acid sequences that will fold into proteins of desired activity

computer graphics use of computers to draw pictures; essential to present protein structures

constant domains parts of antibodies, not containing antigen-binding sites

core of the structure a central grouping of secondary structural elements, usually including the active site, that during evolution retain the same basic topology although showing distortion in their assembly

CRISPR clusters of regularly interspaced short palindromic repeats, a powerful method of genome editing

Critical Assessment of Functional Annotation (CAFA) programme organized to test quality of prediction of protein function

Critical Assessment of Structure Prediction (CASP) programme organized to test quality of prediction of protein structure from amino-acid sequence

cryo-electron microscopy a method for determining structures of proteins and aggregates; recent breakthroughs in this technique have greatly improved the quality of the results

C-terminus the end of a polypeptide chain containing a free carboxyl group

D

database a collection of data on some subject, carefully curated for quality, and equipped with methods of information retrieval

denaturation unfolding of protein from a compact native state to an open state in which interresidue interactions are lost

diauxy two growth phases, upon changing the environmental conditions; for instance, change of nutrient

diffraction scattering of wave by a set of slits, or a crystal, producing a non-uniform pattern

directed evolution creation of unnatural proteins with novel desired features by generating variants and selecting favourable ones

disulphide bridge a covalent bond between the sulphur atoms of two cysteine residues in proteins that approach each other nearby in space in a structure

domain a subunit of a protein that itself independently forms a compact unit

domain recombination a mode of evolution in which proteins form from different combinations, or orders, of domains

dyneins cytoskeletal motor proteins that move along microtubules in cells, carrying cargo

E

EC number an index in the Enzyme Commission classification of protein catalytic functions

Ehlers-Danlos syndrome an inherited disease resulting in weakened connective tissue

electron-density map a measurement of the electron distribution into which a model can be built

electron-transport chain passage of electrons through a series of carriers, driven by the reoxidation of NADH to NAD and $FADH_2$ to FAD, coupled to pumping of protons across the inner membranes of mitochondria, and chloroplasts, to create a transmembrane pH gradient

electrospray ionization (ESI) a method for vaporizing and ionizing proteins, for mass spectrometry

enolase superfamily proteins, with a large variety of functions, homologous to enolase

entropy a thermodynamic quantity reflecting the conformational freedom of a system

Enzyme Commission a project to devise a classification of protein catalytic functions

evolutionary tree a graphical depiction of relationships among species

exon a region of DNA containing information to be translated into part of the amino-acid sequence of a protein

expansion of the genetic code a laboratory technique by which ribosomes can incorporate non-natural amino acids into protein structures

F

feedback inhibition the reduction or suppression of a metabolic pathway in response to adequate amounts of the final product

fibrous protein an elongated, insoluble protein that provides structural support in cells and tissues

G

gel electrophoresis separation on a polyacrylamide surface of proteins in a mixture based on differential mobility in an electric field

Gene Ontology a project to devise a complete classification of protein function

genetic counselling advising prospective parents about the danger to their children of a genetic disease

genetic drift change in allele frequencies in a population *not* the result of selective pressure

genetic engineering artificial modification of genomes

genome the DNA or RNA containing the genetic information in a cell or virus

genomics the study of genomes

Gibbs Free Energy the thermodynamic quantity that determines spontaneity and equilibrium at constant temperature and pressure: a process will be spontaneous if the change in Gibbs Free Energy is negative; and a system will be at equilibrium if it is in the accessible state of minimal Gibbs Free Energy

glycosylation addition of a carbohydrate to the sidechain of a protein

GroEL–GroES a chaperone system in *E. coli* that unfolds misfolded proteins, affording them another chance to fold properly

H

haemoglobin a protein in the red blood cells of humans and other animals, responsible for transport of oxygen and carbon dioxide

heavy chains subunits of antibody molecules

herceptin a humanized monoclonal antibody in clinical use against breast cancer

high-lysine rice genetically-modified rice that contains a higher concentration of lysine than normal strains

hinge motions a type of conformational change in proteins in which two domains remain relatively rigid but change their relative orientation through large changes in conformation in the small numbers of residues linking the domains

His-tag a short string of histidines attached to an end of a polypeptide; useful in purification because of its high affinity for metal ions, including nickel, cobalt, and copper, which can be immobilized in a column

HIV-1 the virus that causes AIDS

homologues, homologous family of proteins proteins related by evolution

homology evolutionary relationship

homology modelling prediction of the structure of a protein from its amino-acid sequence, based on the known structures of related proteins

humanized antibodies genetically engineered antibodies that, in order to reduce antigenicity, contain the antigen-binding site of a non-human antibody grafted into a human antibody

hydrogen bond a weak interaction between two electronegative atoms, mediated by a hydrogen atom

hydrophobic effect a thermodynamically unfavourable interaction of a substance with water; responsible for the low solubility of benzene in water and for the folding of proteins to sequester non-polar sidechains in the protein interior

hypervariable regions regions of antibody molecules that create the antigen-binding site

I

induced fit change in conformation of a protein resulting from ligand binding

insertions and deletions modifications in genomes such that residues are either added or removed

intermediary metabolism the transformations of small molecules in cells, including both synthesis and degradation

intrinsic disorder regions of native states of proteins in which the structure is not unique and well-defined

intron a region in a DNA sequence intervening between regions (exons) that code for part of the amino-acid sequence of a protein. Introns must be 'spliced out' in processing of messenger RNA, before translation

inverse folding problem given a protein structure, to design an amino-acid sequence that will fold up into that structure

inversions a change in a genome involving reversal of direction of a region

ion-exchange chromatography a method of separating proteins on the basis of charge, based on the attraction of a charged protein to material in a column of the opposite charge

isoelectric focussing (IEF) electrophoretic motion of the molecules in a mixture of proteins in a pH gradient; each molecule will move until it arrives at a pH at which its charge is zero, then it will stop moving

isoelectric point (pI) the pH at which the charge on a protein is zero

I-TASSER a very powerful homology-modelling program

K

kinesins cytoskeletal motor proteins that move along microtubules in cells, carrying cargo

L

light chains subunits of antibody molecules

M

main-chain the polypeptide backbone of a protein; the structure exclusive of the sidechains

Marfan syndrome an inherited disease resulting in weakened connective tissue

mass spectrometry a method of identifying proteins by measurement of charge/mass ratios of fragments

matrix-assisted laser desorption/ionization (MALDI) a method of vaporizing and ionizing molecules for mass spectrometry

maximum velocity, V_{max} the maximum initial velocity of an enzyme-catalysed reaction in presence of high concentrations of substrate

messenger RNA RNA containing protein-coding information transcribed from DNA

metabolic pathways network of molecular transformations

Michaelis constant K_M the substrate concentration at which the initial velocity of an enzymatic reaction is half the maximum value observed at high substrate concentration; also the equilibrium constant for dissociation of the enzyme-substrate complex

modular proteins proteins composed of multiple domains

monoclonal antibody a single specific antibody molecule, unlike the heterogeneous mixture produced naturally in response to challenge by an antigen

moonlight for a protein to show a subsidiary function in addition to its main function

motifs subunits of protein structures that recur in many different proteins

mRNA editing alteration of the nucleotide sequence of a messenger RNA, introducing changes from the faithful copy of a region of DNA

multimeric structures molecules composed of more than one protein and/or nucleic acid

N

native conformation the equilibrium conformation into which an amino-acid sequence will fold; compact, usually unique, and biologically active

neonatal screening testing of newborn infant for genetic disease

NMR spectroscopy a method of determining protein structures by measuring excitation and excitation transfer of nuclei

N-terminus the end of a polypeptide chain containing a free amino group

O

one gene, one enzyme hypothesis the idea, proposed by Beadle and Tatum in 1941, that each gene is associated with a single protein

operon a region in bacterial DNA containing genes for several proteins, plus control regions

orthologues two related proteins in different species

oxidation-reduction reaction a reaction in which one molecule transfers an electron to another

P

paralogues two related proteins in the same species

peptide bond the linkage between any two successive amino acids in a protein

phenylketonuria a disease caused by dysfunction of phenylalanine hydroxylase, the enzyme that converts phenylalanine to tyrosine, causing buildup of phenylalanine and toxic by-products

phospholipid bilayer the basis of membrane structure, containing molecules with a charged head and non-polar tail, packed in double sheets, tail-to-tail, with the charges pointing out from both the inside and outside surfaces of the membrane

polypeptide chains polymers of amino acids, connected by peptide bones

post-translational modification chemical changes to residues after protein synthesis

potassium channel a molecule embedded in a cell membrane that specifically allows passage of potassium ions

primary structure the covalent bonds linking amino acids into a polypeptide chain

promoter a region of DNA to which proteins bind to initiate transcription

protein complexes stable interactions between two or more proteins

protein dysfunction failure of a protein to carry out its normal function, generally caused by mutation; can be the cause of disease

protein expression patterns the distribution of amounts of different proteins synthesized, which may vary from (1) tissue to tissue, and (2) under different conditions, such as yeast growing aerobically or anaerobically

protein structure prediction given an amino-acid sequence of a protein, predict the structure into which it will fold

proteomics the study of the nature, amounts, and distribution of proteins in a cell, tissue, or organism

Q

quaternary structure formation of a protein from multiple polypeptide chains, specifying the composition and the structure of the association between subunits

R

regulatory networks an organized set of relationships that respond to signals and control patterns of function or expression

repressor a molecule that interacts with DNA to reduce or even entirely prevent transcription

reverse transcription copying of RNA sequences of a virus into the host DNA

ribosomal frameshifting exception to the general rule that messenger RNA is 'read' by the ribosome as consecutive sets of three nucleotides

ribosome a molecular machine that 'reads' messenger RNA and synthesizes a polypeptide chain according to the codons in the messenger

ribozymes an RNA molecule that has catalytic activity

root-mean-square deviation (r.m.s.d.) a measure of the similarity of two structures

S

Salting in the observation that at low salt concentrations, protein solubility increases with salt concentration

Salting out the observation that at high salt concentrations, protein solubility decreases with salt concentration

Sasisekharan-Ramakrishnan-Ramachandran plot a graph showing the main-chain conformational angles ϕ and ψ of the residues in a protein

scurvy a disease resulting from vitamin C deficiency, weakening connective tissues

secondary structure the distribution in a protein structure of α–helices and β–sheets

sickle-cell disease the result of a mutation in haemoglobin, resulting in polymerization of haemoglobin in the deoxy state, impeding blood flow

sidechains the variable portions of amino acids

signal reception recognition by a cell of the arrival of a molecule, initiating a response

signal-receptor proteins a protein on the surface of a cell to which a signalling molecule binds, triggering an intracellular response

sodium dodecyl sulfate (SDS) a detergent that denatures proteins and coats them with a uniform layer of negative charge

splice variants proteins encoded by the same region of DNA but containing different subsets of the amino-acid sequence specified

spontaneous folding of proteins the achievement by proteins of native states, based on no information other than the amino-acid sequences

Structure-Function Linkage Database (SFLD) a database presenting relationships between amino-acid sequences and functions in protein families

superposition moving two or more structures around in space to bring corresponding points as close together as possible

supersecondary structures compact combinations of secondary structures appearing sequentially in protein structures

synchrotron a particle accelerator producing a very bright X-ray beam, applicable to data collection for protein crystallography

T

tertiary structure the folding up in space of a polypeptide chain

transcriptomics the study of the nature, amounts, and distribution of RNA molecules in a cell, tissue, or organism

translocations a change in DNA sequence in which a region is moved to a different place in the genome sequence

transmembrane channels proteins embedded in membranes that allow and control transport of substances across the membrane

transport proteins a protein responsible for moving a substance around a cell or around the body

twilight zone the region of sequence similarity between 20% and 30% sequence identity; with this degree of similarity it is difficult to decide whether two proteins are related or not

V

Van der Waals forces general cohesive forces between all atoms

variable domains components of antibody molecules, containing the antigen-binding site

variable splicing different choices of exons during processing of messenger RNA, to create several proteins from a single coding region in DNA

voltage-gated potassium channel a channel that permits passage of potassium ions when the voltage across the membrane falls below a threshold

voltage-gated sodium channel a channel that permits passage of sodium ions when the voltage across the membrane falls below a threshold

volume-exclusion chromatography a method of separation of molecules in a mixture, on the basis of molecular volume

X

X-ray crystallography the determination of chemical structures, including proteins, from the diffraction patterns of crystals

INDEX

Notes: Tables and figures are indicated by an italic, *t* and *f* following the page number